Lecture Notes in Mechanical Engineering

For further volumes:
http://www.springer.com/series/11236

About this Series

Lecture Notes in Mechanical Engineering (LNME) publishes the latest developments in Mechanical Engineering—quickly, informally and with high quality. Original research reported in proceedings and post-proceedings represents the core of LNME. Also considered for publication are monographs, contributed volumes and lecture notes of exceptionally high quality and interest. Volumes published in LNME embrace all aspects, subfields and new challenges of mechanical engineering. Topics in the series include:

- Engineering Design
- Machinery and Machine Elements
- Mechanical Structures and Stress Analysis
- Automotive Engineering
- Engine Technology
- Aerospace Technology and Astronautics
- Nanotechnology and Microengineering
- Control, Robotics, Mechatronics
- MEMS
- Theoretical and Applied Mechanics
- Dynamical Systems, Control
- Fluid Mechanics
- Engineering Thermodynamics, Heat and Mass Transfer
- Manufacturing
- Precision Engineering, Instrumentation, Measurement
- Materials Engineering
- Tribology and Surface Technology

Fakher Chaari · Jacek Leśkow
Antonio Napolitano · Andrea Sanchez-Ramirez
Editors

Cyclostationarity: Theory and Methods

 Springer

Editors
Fakher Chaari
Mechanical Engineering Department
National School of Engineers of Sfax
Sfax
Tunisia

Jacek Leśkow
Institute of Mathematics
Cracow University of Technology
Cracow
Poland

Antonio Napolitano
Department of Engineering
University of Napoli "Parthenope"
Napoli
Italy

Andrea Sanchez-Ramirez
Faculty of Engineering Technology
University of Twente
Enschede
The Netherlands

ISSN 2195-4356   ISSN 2195-4364   (electronic)
ISBN 978-3-319-04186-5   ISBN 978-3-319-04187-2   (eBook)
DOI 10.1007/978-3-319-04187-2
Springer Cham Heidelberg New York Dordrecht London

Library of Congress Control Number: 2014930090

Printed on acid-free paper

Springer is part of Springer Science+Business Media (www.springer.com)

Contents

Introduction

In the last decade, the research in signal analysis was dominated by models that encompass nonstationarity as an important feature. The workshop held in Grodek—Poland in February 2013 was dedicated to investigation of cyclostationary signals, that is, signals that exhibit some strict and/or approximate periodicity. The main objective is to highlight the strong interactions between theory and applications of cyclostationary signals with the use of modern statistical tools. In our opinion, the methods like bootstrap, subsampling, or Fraction-of-time (FOT) model approach will revolutionize the signal analysis in many applied engineering areas. A special focus was made on applications in vibro-mechanical signals. Important features of the signals are studied in the time and frequency domains. In the time domain, the relevant objects are first- and second-order moment characteristics, whereas in frequency domain one wants to study spectral density, spectral coherencies, and spectral kurtosis. The proper perspective of studying time and spectral characteristics of nonstationary and repeatable signals is using almost periodically correlated models since this allows proper identification of relevant frequencies, estimation of time characteristics, and more advanced statistical studies.

One has to state very clearly that without a good mathematical and statistical formalism, one can hardly cope with the simplest signal processing procedures as identifying frequencies for a nonstationary or cyclostationary signal. Without a good quantitative model, many practical procedures are heavily dependent on particular choices of experimental designs, sizes of smoothing windows, etc. Therefore, the fundamental research is of utmost importance to provide reliable tools for researchers. An important application of cyclostationary signals is the analysis of mechanical signals generated by a vibrating mechanism. Cyclostationary models are important to perform basic operations on signals in both time and frequency domains. One of the fundamental problems in diagnosis of rotating machine is the identification of significant modulating frequencies that contribute to the cyclostationary nature of the signals. Classical statistical methods for frequency identification in cyclostationary signals were based on the assumption of gaussianity of the signal and on the assumption of some linear structure of the signal. Our aim is to show that there are modern tools available for analyzing cyclostationary signals without the assumption of gaussianity. These methods are based on the ideas of bootstrap, subsampling, and Fraction-of-time (FOT) models.

The book is organized into two parts. Part I is dedicated to pure theory on cyclostationarity.

Applications are presented in Part II including several mechanical systems such as bearings, gears, ... with or without damages.

Part I
Theory of Cyclostationarity

Time-Angle Periodically Correlated Processes

Jérôme Antoni, Dany Abboud and Sophie Baudin

Abstract Cyclostationary processes have now become an essential mathematical representation of vibration and acoustical signals produced by rotating machines. However, to be applicable the approach requires the rotational speed of the machine to be constant, which imposes a limit to several applications. The object of this chapter is to introduce a new class of processes, coined time-angle periodically correlated, which extends second-order cyclostationary processes to varying regimes. Such processes are fully characterized by a time-angle autocorrelation function and its double Fourier transform, the frequency-order spectral correlation. The estimation of the latter quantity is briefly discussed and demonstrated on a real-world vibration signal captured during a run-up.

1 Introduction

During the last two decades, several research works have highlighted the benefits gained from modeling some mechanical signals as cyclostationary processes (Antoniadis and Glossiotis 2001; Antoni 2009; Leśkow 2012; Raad et al. 2008; Randall et al. 2001). This is especially true for vibration or acoustical signals produced by rotating machinery, since they are intrinsically linked to some periodic mechanisms. Some typical examples are vibrations of gears which are essentially first-order cyclostationary (Capdessus et al. 2000), at least as far as modal vibrations are

J. Antoni (✉)
Laboratoire Vibrations Acoustique (LVA), University of Lyon, Villeurbanne, France
e-mail: jerome.antoni@insa-lyon.fr

D. Abboud
Centre Technique des Industries Mécaniques (CETIM), Senlis, France

S. Baudin
Laboratoire de Mecanique des Contacts et des Structures (LaMCoS) UMR5259,
University of Lyon, Villeurbanne, France

F. Chaari et al. (eds.), *Cyclostationarity: Theory and Methods*,
Lecture Notes in Mechanical Engineering, DOI: 10.1007/978-3-319-04187-2_1,
© Springer International Publishing Switzerland 2014

concerned, and vibrations of rolling element bearings, which are essentially second-order cyclostationary (Antoni 2007a; Cioch et al. 2013), also referred to as periodically-correlated. In this context, the cyclostationary framework has been found extremely useful not only for formally describing the signals of interest, but also for solving theoretical problems which were hardly tractable otherwise (Antoni 2009); blind extraction or separation of signals are such examples, but not only. Indeed, as far the detection and extraction of weak repetitive mechanical signatures is of concern—a situation relevant to many types of incipient faults in rotating machines—the cyclostationary approach is difficult to compete with (Borghesani et al. 2013a).

The cyclostationary framework is large enough to include periodic signals, stationary random signals, and some non-stationary random signals with hidden periodicities in the form of periodic modulation of stationary random carriers. This is wide enough to represent most signals from rotating machines, provided that the machines are operating under a "stationary" regime. In the strict sense, this means time-cyclostationarity (i.e. some statistics are periodic with respect to the time variable) is produced only if the rotation speed of the mechanism is either (i) constant or has possible (ii) stationary or (iii) cyclostationary random fluctuations (small enough to maintain a positive speed)—see (Antoni et al. 2004) for a discussion of this point. Unfortunately, the assumption of a stationary regime, even if classical in numerous research works, is limitative in several respects (Borghesani et al. 2013b). One reason is that some rotating machines cannot be locked to a constant regime, such as windturbines or crushers, which are constantly subjected to strongly varying loads. Another reason is that the information of interest (e.g. the mechanical signature of a fault) is sometimes revealed during a transient regime, such as the run-up of an engine. In such scenarii, the cyclostationary approach may fall short.

It is the object of this chapter to propose an extension of the cyclostationary framework to signals that undergo non-stationary speed regimes. This will naturally drive the reader outside the cyclostationary sphere towards a larger class of non-stationary signals. The main difficulty in doing so is to arrive at a formalism that still enjoys enough properties to be of practical usefulness and to permit rigorous statistical definitions (in particular of estimators) (Napolitano 2012). For the sake of simplicity but also of practical interest, only second-order statistics will be considered here; the corresponding processes are commonly referred to as "periodically correlated" (Dehay 1994; Hurd 1989, 1991; Yavorsky et al. 2011).

The rest of the chapter is organized as follows. Section 2 reviews the time and angle description of periodically correlated processes, two alternative views of the same reality which are frequently contrasted when speed fluctuations is of concern. Next, Sect. 3 proposes a definition of a certain class of processes that extend the periodic correlation property to non-stationary regimes. The corresponding processes will be referred to as time-angle periodically correlated. Then, Sect. 4 introduces the order-frequency spectral correlation, which fully describes such processes. Estimation issues are also addressed in the same section. Finally, an example of application on real-life signals is provided in Sect. 5.

2 Time Versus Angle Periodicity

2.1 Problem Statement

When concerned with signals produced by rotating machines, a fundamental issue is to choose between a description in terms of the temporal or the angular variable (it is likely that similar issues arise in other fields of application). The angular variable reflects the rotation of a given shaft of reference (e.g. the primary shaft) in the machinery and, as such, all periodic mechanisms resulting from the rotation of the mechanical elements should be expressed with respect to it. This is referred to as angle-periodicity. Time t and angle θ are related by the rotation speed

$$\dot{\theta} = \frac{d\theta}{dt}, \tag{1}$$

where it is assumed that $\dot{\theta} > 0$ (i.e. there is no backlash). Obviously, if the speed was constant, then time and angle would be linearly related and angle and time-periodicities completely equivalent. When applied to the statistics of a stochastic process, the same conclusion would hold about angle and time-cyclostationarity. However, when speed is not constant—which is the situation addressed in this chapter—the angle-periodicity of a rotating machine still holds, but not its time-periodicity. Yet this does not mean the resulting signals (e.g. the radiated noise or generated vibrations) are angle-cyclostationary in general. The reason is because such signals are related to physical phenomena which are described by time-dependent dynamical characteristics, such as temporal differential equations. Therefore, time-dependence will interfere with angle-periodicity and, ultimately, will make it collapse. Two simple examples of this dilemma are as follows.

(a) *Angle-periodic amplitude modulation*

Let us consider a zero-mean, white, stationary, random process, $\varepsilon(t)$, and an angle-periodic function $p(\theta) = p(\theta + \Theta)$, where Θ stands for the angular period (typically $\Theta = 2\pi$). Then, an elementary model for an angle-periodically modulated process is

$$X(t) = p\left(\theta(t)\right)\varepsilon(t), \tag{2}$$

as would occur for instance when the rotation of a blade modulates a steady flow of fluid. The time-domain autocorrelation function of $X(t)$ reads

$$R_{2X}(t, \tau) \stackrel{def}{=} \mathbb{E}\left\{X(t) X(t - \tau)\right\} = p\left(\theta(t)\right)^2 \sigma_\varepsilon^2 \delta(\tau) \neq R_{2X}(t + T, \tau) \tag{3}$$

with σ_ε^2 the variance of $\varepsilon(t)$. This clearly is not periodic in general since there does not exist a non-zero period T such that $p\left(\theta(t)\right) = p\left(\theta(t + T)\right)$ for all values of time t, except in the particular case with constant speed, $\theta = \Omega t$. Alternatively, when expressed with respect to the angular variable, the autocorrelation function reads

$$R_{2X}(\theta, \phi) \stackrel{def}{=} \mathbb{E}\{X(t(\theta))X(t(\theta - \phi))\} = p(\theta)^2 \sigma_\varepsilon^2 \delta(\phi)\dot\theta \neq R_{2X}(\theta + \Theta, \phi) \tag{4}$$

which again is not periodic since $\dot\theta(\theta) \neq \dot\theta(\theta + \Theta)$ in general.

(b) *Angle-periodic frequency modulation*

Let us now consider a zero-mean, white, stationary, random process, $\varepsilon(t)$, that is input to a time-homogeneous differential equation,

$$\ddot X(t) + c \cdot \dot X(t) + k(\theta(t)) \cdot X(t) = \varepsilon(t) \tag{5}$$

with angle-periodically varying stiffness $k(\theta)$. This provides a simple model to any linear angle-periodic system excited by a stationary force, such as encountered in rotating machines. In this case, the time-domain autocorrelation function of solution $X(t)$ is returned by the implicit equation

$$\left(\frac{\partial^2}{\partial \tau^2} + c\frac{\partial}{\partial \tau} + k(\theta(t + \tau))\right) R_{2X}(t, \tau) = \sigma_\varepsilon^2 \delta(\tau) \Rightarrow R_{2X}(t, \tau) \neq R_{2X}(t + T, \tau) \tag{6}$$

and the angle-domain counterpart by

$$\left(\frac{\partial^2}{\partial \phi^2} + c\frac{\partial}{\partial \phi} + k(\theta)\right) R_{2X}(\theta, \phi) = \sigma_\varepsilon^2 \delta(\phi)\dot\theta \Rightarrow R_{2X}(\theta, \phi) \neq R_X(\theta + \Theta, \phi). \tag{7}$$

For the same reasons as before—i.e. $k(\theta(t + \tau)) \neq k(\theta(t + T + \tau))$ and $\dot\theta(\theta) \neq \dot\theta(\theta + \Theta)$—none of these functions is periodic in general, in either the time or the angular domain.

2.2 Time-Angle Autocorrelation Function

The previous section has demonstrated that neither the time-domain nor the angle-domain representation can lead to the periodic correlation property under non-stationary regime. Indeed, the dilemma of having to choose between one or the other representation has been a subject of controversy for some time. A main contribution of this chapter is to demonstrate that both domains should be used *conjointly* and not oppositely. This leads to the idea of a time-angle autocorrelation function, $R_{2X}(\theta, \tau)$, defined as

$$R_{2X}(\theta, \tau) \stackrel{def}{=} \mathbb{E}\{X(t(\theta))X(t(\theta) - \tau)\}. \tag{8}$$

Note that in the above, the unit of datum θ is radian whereas that of time-lag is second. By doing so, periodicity is locked to the angular position, whereas dynamics is still governed by time as dictated by physical phenomena describing wave propagation. Coming back to the previous two examples, one has

$$R_{2X}(\theta, \tau) = p(\theta)^2 \sigma_\varepsilon^2 \delta(\tau) = R_{2X}(\theta + \Theta, \tau) \tag{9}$$

for the angle-periodic amplitude modulation and

$$\left(\frac{\partial^2}{\partial \tau^2} + c\frac{\partial}{\partial \tau} + k(\theta)\right) R_{2X}(\theta, \tau) = \sigma_\varepsilon^2 \delta(\tau) \Rightarrow R_{2X}(\theta, \tau) = R_{2X}(\theta + \Theta, \tau) \tag{10}$$

for the angle-periodic frequency modulation. In both cases, one now has perfect angle-periodicity. The question now arises to which class of processes does this generalization of cyclostationarity apply?

3 Models of Time-Angle Periodically Correlated Processes

Various generative models to cyclostationary processes have proposed in the literature, some with physical origins (George et al. 1992) and other with grounds in telecommunication engineering (Gardner 1990). Three equivalent models are proposed in this section, which are motivated by their relevance to mechanics. The first two models are inspired from the theory of linear systems, and the last one from the Fourier series description of stochastic processes.

(a) *Angle-varying differential equation*

From a direct generalization of example (a), a time-angle periodically correlated process $X(t)$ may be modeled as the solution of a linear time-domain differential equation with angle-varying coefficients subjected to a stationary random input $\varepsilon(t)$, viz,

$$\left(\sum_k a_k(\theta(t)) \left(\frac{d}{dt}\right)^i\right) X(t) = \varepsilon(t). \tag{11}$$

Such a form should find a direct correspondence with many mechanical models, e.g. with angle-varying stiffness or inertia.

(b) *Angle-varying impulse response*

By denoting $h(\theta(t), t - \tau)$ the impulse response of a system governed by the differential equation (11), at angle θ due to an impulse at time τ, one has directly

$$X(t) = \int h(\theta(t), \tau)\varepsilon(t - \tau)d\tau. \tag{12}$$

Note this is a generalization of the angle-periodic amplitude modulation of example (a), where an infinity of modulations are applied on delayed versions of process $\varepsilon(t)$ and summed up.

(c) *Angle-dependant Fourier series*

The last model results from expanding $h(\theta, t - \tau)$ in Eq. (12) into its Fourier series with respect to θ. It then comes

$$X(t) = \sum_{k \in \mathcal{A}} e^{j2\pi \frac{k\theta}{\Theta}} c_k(t), \tag{13}$$

where, in general, $\{c_k(t); \ k \in \mathcal{A}\}$ is a family of mutually correlated stationary random processes. In words, a time-angle periodically correlated process has a Fourier series whose basis functions are expressed in the angle-domain and whose Fourier coefficients are stationary random processes in the time-domain. This form has the advantage of evidencing an explicit separation of the time and angle variables. Note that the particular case of periodically correlated processes is immediately recovered by setting $\theta = \Omega t$ (Javorskyj et al. 2011). Furthermore, the periodicity of the time-angle autocorrelation function,

$$R_{2X}(\theta, \tau) = R_{2X}(\theta + \Theta, \tau), \tag{14}$$

is readily proved from Eq. (13).

It is important, at this juncture, to insist on the fact that model (13)—or its counterpart (11) or (12)—still lies in a property framework, which is essential to unambiguously define new statistical tools dedicated to the analysis of time-angle periodically correlated processes. This is the object of the next section.

4 The Frequency-Order Spectral Correlation

4.1 Definitions

Just as for periodically correlated processes, the time-angle autocorrelation function is a core statistics from which other specialized tools can be devised, such as the cyclic autocorrelation function, the angle-frequency Wigner-Ville spectrum, and the frequency-order spectral correlation (Antoni 2007b). Only the latter is investigated in this chapter due to its prominence in mechanical applications.

The frequency-order spectral correlation is defined as the following double Fourier transform,

$$S_{2X}(v, f) = \lim_{\Phi \to \infty} \frac{1}{\Phi} \int_{\Phi} \int_{-\infty}^{\infty} R_{2X}(\theta, \tau) e^{-j2\pi v\theta/\Theta} e^{-j2\pi f\tau} d\tau d\theta, \tag{15}$$

where f stands for the classical frequency in Hz and v for "order" (that is the number of events per rotation of Θ radians) without unit. Therefore, $S_{2X}(v, f)$ is able to characterize both the frequency content of phenomena described by time-domain dynamics—i.e. the carrier frequency f—and periodicity linked to angle-domain kinematics—i.e. the modulation frequency v. An alternative definition that is more suited to estimation purposes results from the change of variable $\theta \to t$, viz

$$S_{2X}(v, f) = \lim_{W \to \infty} \frac{1}{W} \int_W \int_{-\infty}^{\infty} R_{2X}(\theta(t), \tau)\dot{\theta}(t)e^{-j2\pi v\theta(t)/\Theta}e^{-j2\pi f\tau}\,d\tau dt, \quad (16)$$

where W is the time interval corresponding to the angular span Φ. A notable particularity of this definition is the weighting by $\dot{\theta}(t)$ which assigns more importance to high-speed regimes. Definition (16) can be shown equal to the following limit

$$S_{2X}(v, f) = \lim_{W \to \infty} \frac{1}{W}\mathbb{E}\left\{\mathcal{F}_W\{X(t)\}^* \mathcal{F}_W\left\{X(t)\dot{\theta}(t)e^{-j2\pi v\theta(t)/\Theta}\right\}\right\}, \quad (17)$$

where $\mathcal{F}_W\{X(t)\}$ denotes the Fourier transform of process $X(t)$ over a time interval of length W. This last form forces one to revisit the interpretation of the frequency-order spectral correlation as the correlation of two different versions of the process, $X(t)$ and $X(t)\dot{\theta}(t)e^{-j2\pi v\theta(t)/\Theta}$, at frequency f. Only when speed is constant is this equivalent to the classical interpretation of correlation between spectral components of the process at two disjoint frequencies, f and $f + \alpha$ with $\alpha = v \cdot \Omega/\Theta$.

4.2 Estimation Issues

The estimation of the frequency-order spectral correlation follows the usual lines of spectral analysis (Dehay 1994). Let us consider the discrete time setting where signal $x(t)$ is replaced by the discrete sequence $x[n]$, $n = 0, \ldots, L - 1$, and where unit sampling frequency is assumed for simplicity. One trivial estimator is returned by the smoothed periodogram, which reads

$$\hat{S}_{2X}(v, f) = \frac{1}{L - 2M} \sum_{n=M}^{L-M-1} \sum_{\tau=-M}^{M} g[\tau]x[n]x[n-\tau]\dot{\theta}[n]e^{-j2\pi v\theta[n]/\Theta}e^{-j2\pi f\tau},$$

$$(18)$$

with $g[\tau] = g[-\tau]$, $g[0] = 1 \geq g[\tau]$, an $2M + 1$ long symmetric lag-window that decreases slowly towards zero as $|\tau|$ approaches M. Less obvious is the averaged periodogram estimator. A simple way to introduce it is to decompose the discrete time signal as

$$x[n] = \sum_{k=0}^{K-1} w_k[n]x[n], \quad (19)$$

where $\{w_k[n] = w[n - k\Delta]; k = 0, \ldots, K - 1\}, 0 < \Delta < N$, is a family of N—long shifted and overlapping data windows such that

$$\sum_{k=0}^{K-1} w_k[n] = 1. \tag{20}$$

Inserting into Eq. (17) and ignoring cross-terms, one arrives at

$$\hat{S}_{2X}(v, f) = \frac{1}{K \|w\|^2} \sum_{k=0}^{K-1} DFT_N \{w_k[n]x[n]\}^* DFT_N \left\{w_k[n]x[n]\dot{\theta}[n]e^{-j2\pi v\theta[n]/\Theta}\right\} \tag{21}$$

where DFT_N stands for the N-long discrete-time Fourier transform and where the division by L has been replaced by $K\|w\|^2$ for proper power calibration (one can readily check that $K\|w\|^2 \sim L$). The importance of meeting condition (20) to avoid "cyclic leakage" was demonstrated in (Antoni 2007b). The averaged periodogram estimator makes possible an important simplification when the speed variation is slow enough as compared to the data window length; in this case,

$$\hat{S}_{2X}(v, f) \simeq \frac{1}{K \|w\|^2} \sum_{k=0}^{K-1} \Omega_k DFT_N \{w_k[n]x[n]\}^* DFT_N \left\{w_k[n]x[n]e^{-j2\pi vn\Omega_k/\Theta}\right\} \tag{22}$$

where Ω_k stands for the average speed value in the time interval $k\Delta \leq n < k\Delta + N$. Equation (22) boils down to a weighted average of products of DFTs, with more weight being assigned to high speed intervals, and a frequency shift $v\Omega_k/\Theta$ that is updated in each time interval k.

Being based on a formal time-angle definition of the autocorrelation function, these estimators show minor differences with and improve upon the heuristic definitions first proposed in (D'Elia et al. 2010).

5 Example of Application

The application of the frequency-order spectral correlation is illustrated to detect the presence of an outer-race fault in a rolling-element bearing. In the particular case addressed here the fault could hardly show up during a constant regime because of the very low load applied to the bearing, so that the machine had to be tested during a run-up by rapidly increasing the rotation speed from 0 to 60 Hz within 25 s. The speed profile and the corresponding vibration signal captured by the accelerometer are shown in Fig. 1.

The characteristics of the rolling element bearing are as follows: ball diameter = 7.94 mm, pitch diameter = 33.50 mm, number of elements = 8. This returns an expected ball-pass-frequency on the outer-race (BPFO) at 3.05 orders, that is 3.05 times the rotation speed. The classical spectral coherence (i.e. power-normalized

(a)

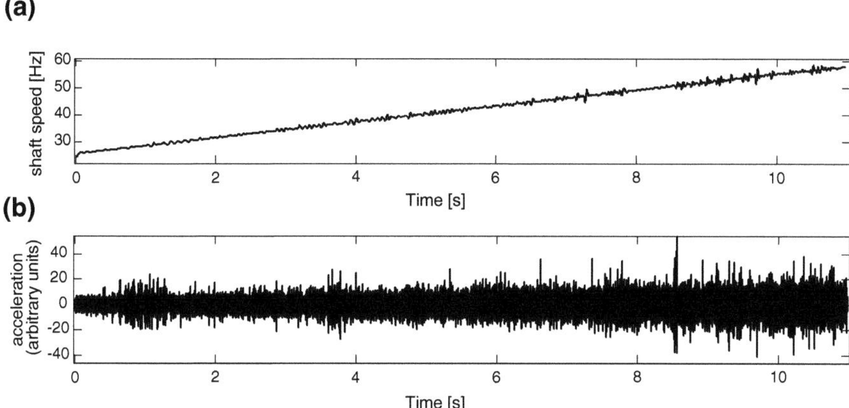

(b)

Fig. 1 **a** Speed profile of the machine and **b** vibration signal

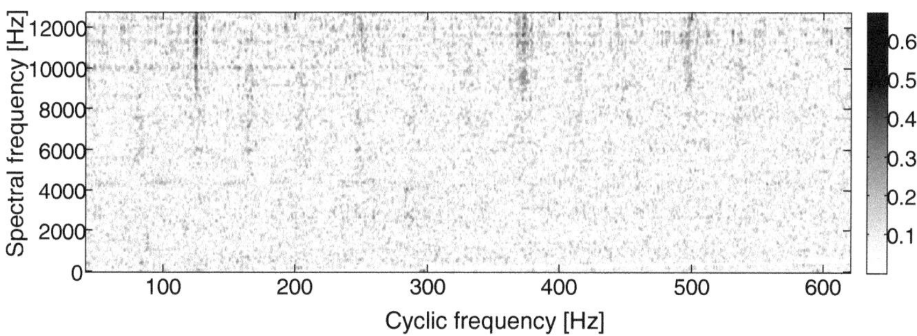

Fig. 2 Classical spectral coherence, $\Delta f = 100\,\mathrm{Hz}$, $\Delta\alpha = 1\,\mathrm{Hz}$

version of the spectral correlation) estimated over the speed interval [25 Hz; 60 Hz] is displayed in Fig. 2 (the averaged periodogram with a hanning window and 66 % overlap was used with imposed spectral resolution $\Delta f = 100\,\mathrm{Hz}$). Spectral lines at the fault frequency and its harmonics are hardly seen above $f = 9\,\mathrm{kHz}$ around $\alpha = 125, 250, 375$ and $500\,\mathrm{Hz}$—which roughly corresponds to 3 times an average speed of 42 Hz—because they are blurred by the speed variation.

Next, the frequency-order spectral coherence was estimated according to formula (21) with similar settings as before. The result shown in Fig. 3 now evidences sharp spectral lines above $f = 6\,\mathrm{kHz}$ at multiple of order 3.05, as expected from an outer-race fault. A slight modulation at order 1 due to input shaft rotation is also noticeable, as sometimes happens with this kind of fault.

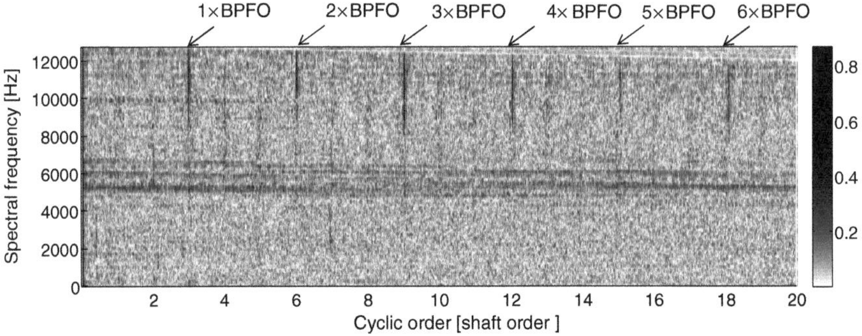

Fig. 3 Frequency-order spectral correlation, $\Delta f = 100\,\text{Hz}$, $\Delta v = 0.02$

6 Conclusion

The generalization of cyclostationary processes to a wider class of non-stationary conditions is a current and challenging topic of research. In particular, the analysis of non-stationary operating conditions of rotating machines is one field where the subject finds considerable interest. This chapter has shown that one way to tackle the issue is to consider the time and angle domains jointly instead of oppositely. This makes possible to extent significantly the applicability of cyclostationary processes. The present work has addressed only the second-order case (i.e. periodically corre-lated processes), but there is no doubt that similar lines can be followed on higher-orders. One limitation of the proposed approach is that it does not assume explicit dependence on the rotation speed (although such dependence is implicit in the rela-tionships between angle and time). This precludes the analysis of speed-dependant modulations, such as resonances due to the passage of critical speeds. Yet, the authors believe the proposed approach is the first step towards the consideration of more gen-eral scenarii, including the latter one.

Appendices

A.1 Proof of Eq. (17)

The key point is to carefully rewrite the interval of integration over τ as $[t - W/2; t + W/2]$ when W grows to infinity. Thus,

$$S_{2X}(v, f) = \lim_{W \to \infty} \frac{1}{W} \int_{-W/2}^{W/2} \int_{t-W/2}^{t+W/2} \mathbb{E}\{X(t)X(t - \tau)\}\dot{\theta}(t)e^{-j2\pi v\theta(t)/\Theta}e^{-j2\pi f\tau}\,d\tau dt$$

$$= \lim_{W \to \infty} \frac{1}{W} \int_{-W/2}^{W/2} \int_{-W/2}^{W/2} \mathbb{E}\{X(t)X(u)\}\dot{\theta}(t)e^{-j2\pi v\theta(t)/\Theta}e^{-j2\pi f(t-u)}\,du\,dt$$

$$= \lim_{W \to \infty} \frac{1}{W}\mathbb{E}\left\{ \int_{-W/2}^{W/2} X(u)e^{j2\pi fu}\,du \int_{-W/2}^{W/2} X(t)\dot{\theta}(t)e^{-j2\pi v\theta(t)/\Theta}e^{-j2\pi ft}\,dt \right\}$$

$$= \lim_{W \to \infty} \frac{1}{W}\mathbb{E}\left\{ \mathcal{F}_W\{X(t)\}^* \, \mathcal{F}_W\left\{X(t)\dot{\theta}(t)e^{-j2\pi v\theta(t)/\Theta}\right\} \right\} \tag{A1}$$

A.2 Symmetric Statistics

A symmetric version of the time-angle autocorrelation function is as follows

$$R_{2X}(\theta, \tau) \stackrel{def}{=} \mathbb{E}\left\{X\left(t(\theta) + \tau/2\right)X\left(t(\theta) - \tau/2\right)\right\}, \tag{A2}$$

which has the advantage of returning an even function of τ. This does not change definition (15) of the frequency-order spectral correlation, yet the counterpart of Eq. (17) requires an approximation. Indeed,

$$S_{2X}(v, f) = \lim_{W \to \infty} \frac{1}{W} \int_{-W/2}^{W/2} \int_{-W+2|t|}^{W-2|t|} \mathbb{E}\left\{X\left(t + \frac{\tau}{2}\right)X\left(t - \frac{\tau}{2}\right)\right\}\dot{\theta}(t)e^{-j2\pi v\frac{\theta(t)}{\Theta}}e^{-j2\pi f\tau}\,d\tau\,dt$$

$$= \lim_{W \to \infty} \frac{1}{W} \int_{-W/2}^{W/2} \int_{-W/2}^{W/2} \mathbb{E}\{X(v)X(u)\}\dot{\theta}\left(\frac{u+v}{2}\right)e^{-j2\pi \frac{v}{\Theta}\theta(\frac{u+v}{2})}e^{-j2\pi f(v-u)}\,dv\,du$$

which cannot be factored into the product of two integrals, unless the speed variation is small enough as compared to the correlation length of the process so that

$$\dot{\theta}\left(\frac{u+v}{2}\right)e^{-j2\pi \frac{v}{\Theta}\theta(\frac{u+v}{2})} \simeq \sqrt{\dot{\theta}(u)\dot{\theta}(v)}e^{-j2\pi \frac{v(\theta(u)+\theta(v))}{2\Theta}} \tag{A3}$$

over the whole domain where $\mathbb{E}\{X(v)X(u)\}$ is significantly different from zero. Therefore,

$$S_{2X}(v, f) \simeq \lim_{W \to \infty} \frac{1}{W}\mathbb{E}\left\{ \mathcal{F}_W\left\{X(t)\dot{\theta}(t)^{\frac{1}{2}}e^{j\pi v\theta(t)/\Theta}\right\}^* \mathcal{F}_W\left\{X(t)\dot{\theta}(t)^{\frac{1}{2}}e^{-j2\pi v\theta(t)/\Theta}\right\} \right\}.$$
$$\tag{A4}$$

Because of the assumption required in (A3), one may prefer the *exact* asymmetric form (17) to the *approximate* symmetric form (A4).

References

Antoniadis I, Glossiotis G (2001) Cyclostationary analysis of rolling-element bearing vibration signals. J Sound Vib 248(5):829–845

Antoni J, Bonnardot F, Raad A, El Badaoui M (2004) Cyclostationary modelling of rotating machine vibration signals. Mech Syst Signal Process 18:1285–1314

Antoni J (2007a) Cyclic spectral analysis of rolling-element bearing signals: facts and fictions. J Sound Vib 304(3–5):497–529

Antoni J (2007b) Cyclic spectral analysis in practice. Mech Syst Signal Process 21(2):597–630

Antoni J (2009) Cyclostationarity by examples. Mech Syst Signal Process 23(4):987–1036

Borghesani P, Pennacchi P, Ricci R, Chatterton S (2013a) Testing second order cyclostationarity in the squared envelope spectrum of non-white vibration signals. Mech Syst Signal Process 40(1):38–55

Borghesani P, Pennacchi P, Chatterton S, Ricci R (2013b) The velocity synchronous discrete fourier transform for order tracking in the field of rotating. Mech Syst Signal Process 40(1):38–55

Capdessus C, Sidahmed M, Lacoume JL (2000) Cyclostationary processes: application in gear faults early diagnosis. Mech Syst Signal Process 14(3):371–385

Cioch W, Knapik O, Leśkow J (2013) Finding a frequency signature for a cyclostationary signal with applications to wheel bearing diagnostics. Mech Syst Signal Process 38(1):55–64

Dehay D (1994) Spectral analysis of the covariance of the almost periodically correlated processes. Stoch Process Appl 50(2):315–330

D'Elia G, Daher D, Antoni J (2010) A novel approach for the cyclo-non-stationary analysis of speed varying signals. In: ISMA2010 international conference on noise and vibration engineering, Leuven

Gardner W (1990) Introduction to random processes, 2nd edn. McGraw-Hill, New York

George V, Gaonkar G, Prasad J, Schrage D (1992) Adequacy of modeling turbulence and related effects on helicopter response. AIAA J 30(6):1468–1479

Hurd H (1989) Nonparametric time series analysis for periodically correlated processes. IEEE Trans InfoTheor 35(2):350–359

Hurd H (1991) Correlation theory of almost periodically correlated processes. J Multivar Anal 37(1):24–45

Javorskyj I, Leśkow J, Kravets I, Isayev I, Gajecka E (2011) Linear filtration methods for statistical analysis of periodically correlated random processes—part ii: harmonic series representation. Signal Process 91(11):251–2506

Leśkow J (2012) Cyclostationarity and resampling for vibroacoustic signals. Acta Physica Polonica A, 121(A): A160–A163

Napolitano A (2012) Generalizations of cyclostationary signal processing: spectral analysis and applications. 1st edn, Wiley, New York

Raad A, Antoni J, Sidahmed M (2008) Indicators of cyclostationarity: theory and application to gear fault monitoring. Mech Syst Signal Process 22(3):574–587

Randall RB, Antoni J, Chobsaard S (2001) The relationship between spectral correlation and envelope analysis for cyclostationary machine signals, application to ball bearing diagnostics. Mech Syst Signal Process 15(5):945–962

Yavorsky I, Kravets I, Mats'Ko I (2011) Spectral analysis of stationary components of periodically correlated random processes. Radioelectron Commun Syst 54(8):451–463

Bootstrap for Maximum Likelihood Estimates of PARMA Coefficients

Anna E. Dudek, Harry Hurd and Wioletta Wójtowicz

Abstract In this chapter we use bootstrap techniques to estimate empirical distributions of parameter estimates for PAR sequences determined by maximum likelihood techniques. The parameters are not the periodic autoregression parameters, but are the coefficients in the Fourier series representing the parameters. We compare two different bootstrap techniques, IID and GSBB, applied to the residuals of the maximum likelihood estimation. The IID method seems a little better, which is not a surprise since the conditions for the GSBB are not completely satisfied. We expect these method to also work satisfactorily for full PARMA estimations, where both PMA and PAR terms are present in the model.

1 Introduction

Let $\{X(t), t \in \mathscr{Z}\}$ be a PARMA (p, q) (periodic autoregressive-moving-average) time series with the known period of the length T i.e.

$$
X_t = \sum_{j=1}^{p} \phi_j(t) X_{t-j} + \sum_{k=1}^{q} \theta_k(t) \xi_{t-k} + \sigma(t) \xi_t, \tag{1}
$$

where $\phi_j(t) = \phi_j(t+T), \theta_k(t) = \theta_k(t+T), \sigma(t) = \sigma(t+T)$ for all $j = 1, \ldots, p$, $k = 1, \ldots, q$ are periodic coefficients, and ξ_t is mean zero white noise with variance equal to one.

A. E. Dudek (✉) · W. Wójtowicz
AGH University of Science and Technology, al. Mickiewicza 30, 30-059 Krakow, Poland
e-mail: aedudek@agh.edu.pl

H. Hurd
Department of Statistics and Operations Research, University of North Carolina at Chapel Hill, Chapel Hill, NC 27599-3260, USA

F. Chaari et al. (eds.), *Cyclostationarity: Theory and Methods*,
Lecture Notes in Mechanical Engineering, DOI: 10.1007/978-3-319-04187-2_2,
© Springer International Publishing Switzerland 2014

Examples of PARMA times series can be found e.g. in Hurd and Miamee (2007).

Several methods have been proposed for the estimation of the parameters in the PARMA model (1). The first seems to have been the method of Maximum Likelihood introduced by Vecchia (1985) and more recently a method using the innovations algorithm introduced by Anderson et al. (2007). We have concentrated on the maximum likelihood method applied not to the parameters themselves $\{\phi_j(t), j = 1, ..., p; \quad \theta_k(t), k = 1, ..., q; \quad \sigma(t)\}$ for $t = 0, 1, ..., T - 1$, but the coefficients in their Fourier transforms. Since, for the maximum likelihood method, the estimates are not expressed directly in terms of the data, computation of estimation error is not straightforward. However, sample distributions of estimates can be computed via bootstrap in a rather straightforward way.

An alternative parametrization that uses Fourier representation was introduced by Jones and Brelsford (1967) to reduce the number of parameters required to represent PARMA model.

$$\phi_j(t) = a_{j,1} + \sum_{m=1}^{\lfloor T/2 \rfloor} a_{j,2m} \cos(2\pi mt/T) + \sum_{m=1}^{\lfloor T/2-1 \rfloor} a_{j,2m+1} \sin(2\pi mt/T), \quad j = 1, ..., p,$$

$$\theta_k(t) = b_{k,1} + \sum_{m=1}^{\lfloor T/2 \rfloor} b_{k,2m} \cos(2\pi mt/T) + \sum_{m=1}^{\lfloor T/2-1 \rfloor} b_{k,2m+1} \sin(2\pi mt/T), \quad k = 0, ..., q,$$

where $\theta_0(t) = \sigma(t)$. The transformation is one-to-one when parameters are unrestricted but provides a simple, and sometimes physically motivated, way to reduce the number of parameters by restricting the the number of frequencies in the Fourier series of $\phi_j(t), \theta_k(t), \sigma(t)$. Maximum likelihood estimates are made of the restricted model parameters (some subset of the unrestricted $\{a_{j,m}, b_{k,m}\}$) and finally the solution can be transformed to $\{\phi_j(t), \theta_k(t)\}$.

2 Bootstrap Methods

In the sequel we propose to use two different bootstrap methods to obtain the confidence intervals for the parameters $\{a_{j,m}, b_{k,m}\}$. The main reason that bootstrap is of interest here is that estimates (at least the maximum likelihood estimates) of the parameters are made indirectly i.e. by maximizing a likelihood calculation, not by a direct expression of the data. The first considered method is based on the idea of bootstrap of residuals for ARMA time series (for more details see e.g. Lahiri 2003). The latter one is using the Generalized Seasonal Block Bootstrap (GSBB) of Dudek et al. (2014). Below we describe both techniques.

Let $X(1), ..., X(n)$ be a sample from PARMA (p,q) time series. To apply any of bootstrap algorithms first the estimates of $\{a_{j,m}, b_{k,m}\}$ coefficients need to be calculated. As a result we get the residuals $\widehat{\varepsilon}_i$ for $i = 1, ..., N$, where $N = n - (p + q)$.

Since for ARMA time series to generate the valid approximation of the asymptotic distribution the residuals need to be centered, we expect that PARMA model also requires this condition. Thus, we define the centered residuals by

$$\tilde{\varepsilon}_i = \widehat{\varepsilon}_i - \bar{\varepsilon}_N, \tag{2}$$

where $\bar{\varepsilon}_N = (N)^{-1} \sum_{i=1}^{N} \widehat{\varepsilon}_i$.

Method 1-IID bootstrap:

1. For $i = 1, \ldots, n$ let

$$\varepsilon_i^{*(1)} = \tilde{\varepsilon}_{k_i},$$

   where k_i is iid from a discrete uniform distribution

$$P\left(k_i = s\right) = \frac{1}{N} \quad \text{for} \quad s = 1, \ldots, N.$$

2. Joining selected residuals we get the bootstrap sample $\left(\varepsilon_1^{*(1)}, \ldots, \varepsilon_n^{*(1)}\right)$.

Method 1 assumes the the residuals $\tilde{\varepsilon}_i$ $(i = 1, \ldots, N)$ are at least approximately independent, so selecting each of them separately we do not destroy the dependence structure in the sample. But in practical cases this condition may not hold. For example one may choose to fit the model of lower rank that the true one. Then, the residuals are no longer independent. Moreover, they may reflect the periodic structure of the original data.

GSBB is the new block bootstrap technique for periodic data. It is the generalization of two known block bootstrap methods i.e. the Seasonal Block Bootstrap of Politis (Politis 2001) and the Periodic Block Bootstrap of Chan et al. (2004). Dudek et al. (2014) used it for the overall mean and the seasonal means of the periodically correlated (PC) time series. Moreover, they showed GSBB consistency for triangular row-wice periodically correlated arrays with growing period. The wide spectrum of possible GSBB applications encouraged us to apply it in the considered problem.

To simplify the presentation of GSBB algorithm we assume that the sample size n is an integer multiple of the block length b $(n = bl)$ and also is an integer multiple of of the period length T $(n = wT)$. Each of these conditions can be easily omitted (for more details see Dudek et al. 2014). Moreover, we present the circular version of GSBB i.e. we treat the sample as wrapped on the circle. Whenever the index t of any chosen observation is greater than N we take $t - N$ instead.

Method 2-GSBB:

1. Choose a (positive) integer block size $b(<N)$.
2. For $t = 1, b + 1, 2b + 1, \ldots, (l - 1)b + 1$, let

$$\left(\varepsilon_t^{*(2)}, \varepsilon_{t+1}^{*(2)}, \ldots, \varepsilon_{t+b-1}^{*(2)}\right) = \left(\widehat{\varepsilon}_{k_t}, \widehat{\varepsilon}_{k_t+1}, \ldots, \widehat{\varepsilon}_{k_t+b-1}\right),$$

where k_t is iid from a discrete uniform distribution

$$P\left(k_t = t + vT\right) = \frac{1}{w} \quad \text{for} \quad v = 0, 1, \ldots, w - 1.$$

Since we consider the circular version of GSBB, when $t + vd > N$ we take the shifted observations $t + vT - N$.

3. Join the l blocks $(\tilde{\varepsilon}_{k_t}, \tilde{\varepsilon}_{k_t+1}, \ldots, \tilde{\varepsilon}_{k_t+b-1})$ to get the bootstrap sample.

Finally, the bootstrap version $\left(X_1^{*(j)}, \ldots, X_n^{*(j)}\right)$ of the original sample $(X_1, \ldots, X_n)$ is obtained using the estimates of $\{\phi_j(t), \theta_k(t)\}$ and the bootstrap error variables $\left(\varepsilon_1^{*(j)}, \ldots, \varepsilon_n^{*(j)}\right)$. The superscript j denotes the chosen bootstrap method. It is equal to 1 for IID bootstrap and 2 for GSBB.

In the next section we present some simulation study results in which we construct the bootstrap pointwise equal-tailed confidence interval for coefficients $\{a_{j,m}, b_{k,m}\}$. The actual coverage probabilities (ACPs) are calculated to compare the performance of both considered bootstrap algorithms. Although, we do not have any theoretical results confirming the consistency of the proposed bootstrap methods, the preliminary simulation results indicate the validity of our procedures.

3 Simulation Study

Our aim is to check the performance of the proposed bootstrap algorithms in the problem od estimating confidence intervals for PARMA model coefficients. In this section we consider a few examples of PARMA time series and calculate the bootstrap equal-tailed pointwise confidence intervals for the coefficients $\{a_{j,m}, b_{k,m}\}$ for $j = 1, \ldots, p$, $k = 0, \ldots, q$ and $m = 1, \ldots, T$. In our study we use procedures first implemented by Hurd (2007) and now available as R package 'perARMA' (Comprehensive R Archive Network reference Dudek et al. 2013).

To reduce the number of parameters that needs to be estimated and decrease the time of computation we restricted our study only to PAR time series. The following examples are considered:

PAR2: the nonzero coefficient are $a_{1,1} = 0.8$, $a_{1,2} = 0.3$, $a_{2,1} = -0.4$ and $b_{0,1} = 1$;

PAR1: the nonzero coefficient are $a_{1,1} = 0.8$, $a_{1,2} = 0.3$ and $b_{0,1} = b_{0,2} = -0.5$.

Note that **PAR2** model has the constant $\sigma(t)$ function (equal to 1) in contrary to the **PAR1** case, where $\sigma(t)$ is periodic. The names **PAR2** and **PAR1** indicate that these are PAR(1) and PAR(2) time series, respectively. This particular choice was caused by the fact that we wanted to restrict the number of parameters and simultaneously investigate the influence of function $\sigma(t)$ on our results.

Unfortunately, $\sigma(t)$ is not the only important factor. Much bigger impact can have the choice of the model fitted to the data. Each practitioner will decide to take

Table 1 Actual coverage probabilities for simulated **PAR2** series

Method	n	ACP									
		E1							E2		
		$a_{1,1}$ (%)	$a_{1,2}$ (%)	$a_{1,3}$ (%)	$a_{2,1}$ (%)	$a_{2,2}$ (%)	$b_{0,1}$ (%)	$b_{0,2}$ (%)	$a_{1,1}$ (%)	$a_{2,1}$ (%)	$b_{0,1}$ (%)
IID	120	92.2	92.2	95.2	92.0	93.4	92.2	93.2	89.6	93.4	89.4
Bootstrap	240	94.4	94.0	93.6	93.6	94.0	94.4	91.8	87.2	92.0	82.6
GSBB	120	89.0	91.2	89.4	87.0	91.0	86.8	87.0	89.2	91.8	84.8
	240	87.8	89.4	89.6	88.8	89.8	91.2	88.2	89.6	91.6	84.6

Columns 1–7 refer to **E1** case and 8–10 to **E2**. Rows 1–2 and 3–4 contain results for IID bootstrap and GSBB, respectively. For both methods ACPs for two sample sizes $n = 120$ and $n = 240$ are presented

the model of lower order if according to some criteria it is comparable to more complicated one. Having this in mind we decided consider the following cases

E1: **PAR2** estimating 7 coefficients i.e. $a_{1,1}, a_{1,2}, a_{1,3}, a_{2,1}, a_{2,2}$ and $b_{0,1}, b_{0,2}$;
E2: **PAR2** estimating 3 coefficients i.e. $a_{1,1}, a_{1,2}$ and $b_{0,1}$;
E3: **PAR1** estimating 6 coefficients i.e. $a_{1,1}, a_{1,2}, a_{1,3}$ and $b_{0,1}, b_{0,2}, b_{0,3}$;
E4: **PAR1** estimating 2 coefficients i.e. $a_{1,1}$ and $b_{0,1}$.

As a result, in **E1** and **E3** cases we estimate more coefficient than are nonzero in reality, while in **E2** and **E4** we always have one less coefficient of each type.

To simulate **E1-E4** 'makeparma' procedure provided by 'perARMA' package was used. This function enables to construct a PARMA type sequence of required length for inputed matrices of coefficients. Two different sample lengths n were taken 120 and 240. As presented approach is based on Fourier representation of model coefficients, we use also 'ab2phth' and 'phth2ab' procedures that enable to transform matrices of coefficients to their Fourier representation and conversely. For each simulated series we fit PAR model using 'parmaf' procedure. The function returns estimates of parameters $\{a_{j,m}, b_{k,m}\}$ as well as series of residuals of fitted model. Next for residuals we apply one of two proposed bootstrap method: IID bootstrap or GSBB. The number of generated bootstrap samples B was 300. In a case of GSBB method we also need to comment the choice of block length b. Since so far, there is no method of optimal block length choice we decided to take $b = \lfloor \sqrt[3]{n} \rfloor$ and $b = T$. The period length T is equal to 12. Taking $b = T$ we wanted to check how the performance of GSBB changes when the longer block is taken. Moreover, $b = T$ is a case when GSBB reduce to SBB. Since the results in both cases were comparable in the sequel we only discuss $b = \lfloor \sqrt[3]{n} \rfloor$ case. Finally, to calculate the bootstrap equal-tailed pointwise confidence intervals bootstrap version of coefficients $\{a_{j,m}, b_{k,m}\}$ were calculated (using 'makeparma' function). The 95 % confidence level was taken. The whole procedure was repeated 500 times and the ACPs were calculated. The results for **E1-E2** are presented in Tables 1 and 2 and for **E3-E4** in Tables 3 and 4.

Table 2 Average lengths of confidence intervals for **PAR2** model

Method	n	Average length of CI									
		E1							E2		
		$a_{1,1}$	$a_{1,2}$	$a_{1,3}$	$a_{2,1}$	$a_{2,2}$	$b_{0,1}$	$b_{0,2}$	$a_{1,1}$	$a_{2,1}$	$b_{0,1}$
IID bootstrap	120	0.3323	0.4693	0.3862	0.3247	0.4796	0.2403	0.3552	0.3332	0.3282	0.2531
	240	0.2323	0.3280	0.2667	0.2314	0.3330	0.1734	0.2498	0.2349	0.2328	0.1828
GSBB	120	0.3127	0.4472	0.3656	0.3179	0.4627	0.2289	0.3403	0.3214	0.3106	0.2515
	240	0.2218	0.3187	0.2566	0.2265	0.3277	0.1658	0.2479	0.2389	0.2501	0.1863

Columns of values 1–7 refer to **E1** case and 8–10 to **E2**. Rows 1–2 and 3–4 contain results for IID bootstrap and GSBB, respectively. For both methods ACPs for two sample sizes $n = 120$ and $n = 240$ are presented

Table 3 Actual coverage probabilities for simulated **PAR2** series

Method	n	ACP						E2	
		E1						E2	
		$a_{1,1}$ (%)	$a_{1,2}$ (%)	$a_{1,3}$ (%)	$b_{0,1}$ (%)	$b_{0,2}$ (%)	$b_{0,3}$ (%)	$a_{1,1}$ (%)	$b_{0,1}$ (%)
IID bootstrap	120	88.6	90.2	92.8	59.4	70.4	94.8	59.0	24.0
	240	90.0	90.4	90.6	78.6	70.4	99.0	36.2	5.8
GSBB	120	87.0	88.0	88.0	72.4	80.8	84.4	58.6	21.0
	240	86.9	87.3	88.3	72.7	82.6	88.1	31.2	4.4

Columns 1–6 refer to **E3** case and 7–8 to **E4**. Rows 1–2 and 3–4 contain results for IID bootstrap and GSBB, respectively. For both methods ACPs for two sample sizes $n = 120$ and $n = 240$ are presented

Table 4 Average lengths of confidence intervals for **PAR2** model

Method	n	ACP						E4	
		E3						E4	
		$a_{1,1}$	$a_{1,2}$	$a_{1,3}$	$b_{0,1}$	$b_{0,2}$	$b_{0,3}$	$a_{1,1}$	$b_{0,1}$
IID bootstrap	120	0.3343	0.3308	0.4965	0.3186	0.4782	0.3407	0.1973	0.2278
	240	0.2076	0.2067	0.3181	0.2010	0.2901	0.1863	0.1334	0.1648
GSBB	120	0.3276	0.3358	0.4772	0.2747	0.4141	0.2369	0.1852	0.2056
	240	0.3248	0.3287	0.4680	0.2735	0.4116	0.2411	0.1210	0.1589

Columns of values 1–6 refer to **E3** case and 7–8 to **E4**. Rows 1–2 and 3–4 contain results for IID bootstrap and GSBB, respectively. For both methods ACPs for two sample sizes $n = 120$ and $n = 240$ are presented

When the number of estimated parameters is big enough i.e. in **E1** case IDD bootstrap definitely outperforms GSBB. In fact it is exactly what one may expect as in these examples the residuals are approximately independent and IID method is the most appropriate. The ACPs for all coefficients are about 2–3% lower than the nominal confidence level independently on the sample size n. For GSBB the corresponding values are even 6% lower. Similar conclusions can be taken in **E3** case for a type parameters, although all ACPs are definitely lower than for **E1**. Surprisingly, IID bootstrap is not working well for b type coefficients.

For $b_{0,1}$ and $b_{0,2}$ the ACPs are about 35 % lower than the nominal coverage probability for $n = 120$ and about 15–25 % for $n = 240$. Let us recall that $b_{0,1}$ and $b_{0,2}$ were the nonzero coefficients. Additionally, for $b_{0,3}$ which in reality is equal to zero IDD bootstrap seems to produce too wide confidence intervals. For $n = 120$ the ACP is almost perfectly equal to 95 % but for $n = 240$ it is close to 1. On the other hand, GSBB provides constantly too low ACPs independently on n, but they are higher compering to IID bootstrap and seem to converge slowly to the nominal coverage probability.

Finally, **E2** and **E4** provide the evidence how destructive influence of the too small set of estimated parameters can be. In fact the performance of the both bootstrap techniques is good for **E2** and the differences between those methods are small. **E2** is a case, where the shocks are constant. The ACPs are similar comparing to **E1** for coefficient of a type and decrease about 5 % for $b_{0,1}$. The noticeable problems appear in **E4** example. For both methods and $n = 120$ the ACPs are very low to became extremely small for $n = 240$. This may indicate that bootstrap is inconsistent in this problem. Let us recall that estimating only $b_{0,1}$ we treat the rest of bs as zeros, which means that $\sigma(t)$ is a constat function. As a result the residuals are definitely dependent.

Although we are aware that we did not provide any theoretical confirmation of validity of the bootstrap methods in the considered problem, the simulation study results seem to be very encouraging. They indicate the consistency of bootstrap. Moreover, probably the practitioner will not be able to use the universal method independently on the PARMA series structure. IDD bootstrap seems to be the best choice when the shocks are constant, while block bootstrap is more appropriate in the opposite case. Additionally, one needs to be extremely careful choosing the size of parameters set that need to be estimated. Despite the longer time of computation, the larger set should be taken to avoid the bootstrap inconsistency.

Acknowledgments Research of Anna Dudek was partially supported by the Polish Ministry of Science and Higher Education and AGH local grant.

References

Anderson PL, Tesfaye YG, Meerschaert MM (2007) Fourier-PARMA models and their application to river flows. J Hydrol Eng 12(5):462–472

Chan V, Lahiri SN, Meeker WQ (2004) Block bootstrap estimation of the distribution of cumulative outdoor degradation. Technometrics 46:215–224

Dudek AE, Hurd H, Wójtowicz W (2013) PerARMA: pacakge for periodic time series analysis, R package version 1.5. http://cran.r-project.org/web/packages/perARMA/

Dudek AE, Leśkow J, Politis DN, Paparoditis E (2014) A generalized block bootstrap for seasonal time series, article first published online: 27 NOV 2013, doi:10.1002/jtsa.12053

Hurd HL, Miamee AG (2007) Periodically correlated random sequences: spectral. Theory and practice. Wiley, New York

Jones R, Brelsford W (1967) Time series with periodic structure. Biometrika 54:403–408

Lahiri SN (2003) Resampling methods for dependent data. Springer, New York

Politis DN (2001) Resampling time series with seasonal components, in Frontiers in data mining and bioinformatics. In: Proceedings of the 33rd symposium on the interface of computing science and statistics. Orange County, California, 13–17 June, pp 619–621

Vecchia AV (1985) Maximum likelihood estimation for periodic autoregressive moving average models. Technometrics 27:375–384

EM-Based Inference for Cyclostationary Time Series with Missing Observations

Christiana Drake, Oskar Knapik and Jacek Leśkow

Abstract Periodically correlated (or cyclostationary) time series are becoming more and more popular in many areas (see Gardner et al. 2006). However, in many practical situations data that can be modeled with such time series is incomplete. Some preliminary results on that problem have been presented in a previous work by Drake et al. (2013) by the authors. In this chapter we propose a new ECM-type algorithm based on conditional likelihood and profile likelihood to extend estimation to the case when observations are missing completely at random.

1 Second-Order Cyclostationary Time Series

Let us assume that a zero-mean second-order cyclostationary time series $\{y_t, t \in N\}$ is observed. More precisely, if $B_Y(t, \tau) = Cov\,(y_t, y_{t+\tau})$ denotes the autocovariance function of $\{y_t\}$, then this function is periodic in t.

Some classes of second-order periodically correlated time series can be approximated as amplitude-modulated time series in the following form:

$$y_t = x_t \cdot c_t, \tag{1}$$

C. Drake
Department of Statistics at University of California at Davis, Davis, USA
e-mail: drake@wald.ucdavis.edu

O. Knapik (✉)
CREATES, Department of Economics and Business at Aarhus University, Aarhus, Denmark
e-mail: oknapik@creates.au.dk

J. Leśkow
Institute of Mathematics at Cracow University of Technology, Kracow, Poland
e-mail: jleskow@pk.edu.pl

F. Chaari et al. (eds.), *Cyclostationarity: Theory and Methods,*
Lecture Notes in Mechanical Engineering, DOI: 10.1007/978-3-319-04187-2_3,
© Springer International Publishing Switzerland 2014

In this model (1) it is assumed that:

AS1 The deterministic sequence $c_t > 0$ $\forall t \in \{1, \ldots, T\}$.

AS2 x_t is a zero-mean stationary Gaussian sequence with a bounded and continuous spectral density.

Assumption (**AS1**) assures that the distribution of y_t is not degenerate at zero.

In this research, we assume that $\{x_t\}$ is an $AR(p)$ process:

$$x_t - \sum_{i=1}^{p} \phi_i x_{t-i} = \varepsilon_t \tag{2}$$

where $\{\varepsilon_t\}$ is a sequence of independent Gaussian zero-mean random variables with finite variance σ^2.

The deterministic sequence c_t is assumed to be a known, periodic function of a finite dimensional unknown parameter vector $\lambda = (\lambda_1, \ldots, \lambda_m)$. A possible representation of such a periodic function c_t is given by:

$$c_t = \exp\left(\sum_{j=1}^{k} \left(\lambda_{1j}\cos\omega_j t + \lambda_{2j}\sin\omega_j t\right)\right), \quad t = 1, \ldots, T. \tag{3}$$

We will assume that k is known, all frequencies are known and of the form $\omega_j \in (0, \pi]$, $i = 1, \ldots, k$: $\omega_j = \frac{2\pi r}{P}$ for some $r = 1, \ldots, \frac{P^*}{2}$, where $P^* = P - 1$ for P odd, $P^* = P$ for P even and P is a known period. Our model parameters are identifiable because the c_t are uniquely determined and there is no scale ambiguity.

Missing data analysis of cyclostationary time series occurs in many time series such as economic time series, mechanical signals and also ocean signals (Stefanakos and Athanassoulis 2001). The aim of this chapter is to provide ECM-based statistical procedures in the situation when complete observation of the periodically correlated time series is not available and data is missing at random. Stefanakos and Athanassoulis (2001) proposed some nonstatistical approach to dealing with missing data. Niu (1996) proposed integration over the missing observations in the case when λ is known. In Drake et al. (2013) the authors have presented a preliminary statistical approach to estimation with missing data. This chapter is a continuation of the previous work and proposes to use ECM-type algorithms adapted to periodic processes to handle missing data problems in cyclostationary models.

This chapter is divided into four sections. Section 2 presents results on conditional likelihood-based inference for cyclostationary time series. Section 3 presents two ECM type algorithm developed and Sect. 4 presents simulations studies investigating the convergence properties of the method and a comparison to the maximum likelihood estimators for the complete data model.

2 Likelihood Based Inference for Cyclostationary Time Series

Likelihood based inference for our model (1) can be conducted in two ways using full likelihood or conditional likelihood function. The purpose of this section is to introduce conditional likelihood function approach while the full likelihood approach was presented in our previous chapter (see Drake et al. 2013). Relevant references corresponding to likelihood-based inference can be found in Hamilton (1994).

In order to present the form of conditional likelihood function and subsequent transformations we will need the following notation. In the sequel the matrix C_{T-p} is an $(T-p) \times (T-p)$ diagonal matrix with diagonal vector $(c_T, \ldots, c_{p+1})$. The symbol $./$ denotes the element by element matrix division. For example let $A = [a_{ij}]$ and $B = [b_{ij}]$ denote two matrices of the same dimension, then $A./B = [a_{ij}/b_{ij}]$. Let also $\theta = (\phi, \lambda, \sigma^2)^T$ be the vector of all unknown parameters and suppose that we have $T = n + p$ observations.

We have the following

Proposition 2.1.   The log-likelihood for a vector of observations $y = (y_T, \ldots, y_{p+1})^T$ corresponding to the model (1) conditional on the first p observations is given by:

$$
\begin{aligned}
l_c(\theta) &= \log f_{y_T,\ldots,y_{p+1}|y_p,\ldots,y_1}\left(y_T, \ldots, y_{p+1} \,\middle|\, y_p, \ldots, y_1 \,;\, \theta\right) \\
&= -\log\left(\left|det(C_{T-p})\right|\right) + \log f_{x_T,\ldots,x_{p+1}|x_p,\ldots,x_1}\left(\tfrac{y_T}{c_T}, \ldots, \tfrac{y_{p+1}}{c_{p+1}} \,\middle|\, \tfrac{y_p}{c_p}, \ldots, \tfrac{y_1}{c_1} \,;\, \theta\right) \\
&= -\log\left(\left|det(C_{T-p})\right|\right) - \tfrac{T-p}{2}\log(2\pi) - \tfrac{T-p}{2}\log\left(\sigma^2\right) + \\
&\quad - \sum_{t=p+1}^{T} \tfrac{1}{\sigma^2}\left(\tfrac{y_t}{c_t} - \sum_{i=1}^{p}\phi_i \tfrac{y_{t-i}}{c_{t-i}}\right)^2.
\end{aligned}
$$

(4)

For the conditional log likelihood function (4) we obtain the following results.

The form of the conditional maximum likelihood estimate of σ^2 given ϕ and λ is

$$
\hat{\sigma}^2_{ML} = \frac{1}{T-p}\sum_{t=p+1}^{T}\left(\frac{y_t}{c_t} - \sum_{i=1}^{p}\phi_i\frac{y_{t-i}}{c_{t-i}}\right)^2. \tag{5}
$$

The form of the conditional maximum likelihood estimate of ϕ given σ^2 and λ is

$$
\hat{\phi}_{ML} = \left[(Y./C)^T(Y./C)\right]^{-1}(Y./C)^T(y./c), \tag{6}
$$

where $c = (c_{p+1}, \ldots, c_T)^T$, $y = (y_{p+1}, \ldots, y_T)^T$, Y and C are $(T-p) \times p$ Toeplitz matrices

$$
Y = \begin{pmatrix} y_p & \cdots & y_1 \\ \vdots & \ddots & \vdots \\ y_{T-1} & \cdots & y_{T-p} \end{pmatrix} \,;\, C = \begin{pmatrix} c_p & \cdots & c_1 \\ \vdots & \ddots & \vdots \\ c_{T-1} & \cdots & c_{T-p} \end{pmatrix}. \tag{7}
$$

The conditional maximum likelihood estimate of λ, for fixed σ^2 and ϕ is given by

$$\hat{\lambda}_{ML} = \arg\max_{\lambda} l_c(\lambda), \tag{8}$$

where

$$l_c(\lambda) = -\log\left(\left|c_{p+1} \cdot \ldots \cdot c_T\right|\right) - \frac{T-p}{2}\log\left(\frac{1}{T-p}(y./c)^T P_1^*(y./c)\right), \tag{9}$$

where P_1 and P_1^* are projection matrices defined as

$$P_1 = (Y./C)\left[(Y./C)^T(Y./C)\right]^{-1}(Y./C)^T, P_1^* = I - P_1, \tag{10}$$

where I is the $(T-p)$-dimensional identity matrix. Proof of this result is presented in the Appendix at the end of this chapter.

From Proposition 2.1. it is clear that the most difficult task is to obtain λ_{ML}. In order to bypass this difficulty we will apply the following step-wise estimation procedure using a profile likelihood approach. Under (AS1), (AS2) the profile likelihood-based estimates for the model under consideration are consistent. The conditional likelihood in this case belongs to the exponential family. For the details see Theorem 9.8 by Pawitan (2001).

The estimates are given by:

Step 1: Estimate λ using formula for $\hat{\lambda}_{ML}$.

$$\hat{\lambda}_{ML} = \arg\max_{\lambda \in \Lambda} l_c\left(y, \lambda, \hat{\phi}_{ML}, \hat{\sigma}^2_{ML}\right).$$

Step 2: Use $\hat{\lambda}_{ML}$ from the previous step to obtain ML estimates of ϕ.
Step 3: Get σ^2 using $\hat{\lambda}_{ML}$ and $\hat{\phi}_{ML}$.

The following example where $\{x_t\}$ is an $AR(1)$ will illustrate the algorithm.

Example 1. The steps for this process are as follows:

Step 1: Obtain $\hat{\lambda}_{ML}$ as

$$\begin{aligned}
\hat{\lambda}_{ML} &= \arg\max_{\lambda \in \Lambda} l_c\left(y, \lambda, \hat{\phi}_{ML}, \hat{\sigma}^2_{ML}\right) \\
&= \arg\max_{\lambda \in \Lambda} \left(term1(\lambda) + term2(\lambda)\right).
\end{aligned} \tag{11}$$

where
$$term1 = -\log\left(\left|det(C_{T-1})\right|\right),$$
$$term2 = -\frac{T-1}{2}\log\left(\frac{1}{T-1}\sum_{t=2}^{T}\left(\frac{y_t}{c_t} - \hat{\phi}_{ML}\frac{y_{t-1}}{c_{t-1}}\right)^2\right).$$

Step 2: Using the formula from Step 1, taking $\lambda = \hat{\lambda}_{ML}$ one obtains

$$\hat{\phi}_{ML} = \sum_{t=2}^{T} \left(\frac{y_t}{c_t\left(\hat{\lambda}_{ML}\right)} \frac{y_{t-1}}{c_{t-1}\left(\hat{\lambda}_{ML}\right)} \right) \Big/ \sum_{t=2}^{T} \left(\frac{y_{t-1}}{c_{t-1}\left(\hat{\lambda}_{ML}\right)} \right)^2 .$$

Step 3: Using ML estimates of λ and ϕ one obtains

$$\hat{\sigma}_{ML}^2 = \frac{1}{T-1} \sum_{t=2}^{T} \left(\frac{y_t}{c_t\left(\hat{\lambda}_{ML}\right)} - \hat{\phi}_{ML} \frac{y_{t-1}}{c_{t-1}\left(\hat{\lambda}_{ML}\right)} \right)^2 .$$

In our proposed algorithm we take advantage of the linear model theory with closed and known forms of conditional maximum likelihood estimators of ϕ and σ^2 given other parameters for $AR(p)$ processes. By profiling out the "nuisance parameters" σ^2 and ϕ we reduce the dimension of the nonlinear optimization problem by $p + 1$. Additional advantages combining conditional likelihood and profile likelihood estimation will be presented in the section on ECM-type inference where the new algorithm will be presented.

3 ECM-Type Algorithms in Likelihood-Based Inference for Second-Order Cyclostationary Time Series with Missing Observations

The expectation-maximization (EM) algorithm is an iterative procedure that can be used for computing the maximum likelihood estimator for data sets with missing observations (Dempster et al. 1977). For the model under consideration, there are several possible approaches to the EM algorithm. It can be used with the full likelihood function which in this case is a well-known problem of inference for incomplete data within the multivariate normal distribution (Schaffer 1997). The conditional likelihood function leads to a modified EM algorithm in the linear regression setting with normal errors. The structure of the model imposes a special EM algorithm to estimate our model with incomplete time data. The algorithms are relatively easy to implement but are not trivial to describe.

The conditional maximum likelihood approach enables the use of the ECM (expectation conditional maximization) algorithm which replaces the M-step of the EM algorithm by several computationally simpler conditional maximization (CM) steps. A CM-step may be in closed form or it might itself require iteration. However the CM maximizations are over smaller dimensional spaces, so they are simpler, faster and more stable than the corresponding full maximization called for in the M-step of the EM algorithm, especially when iterations are required. The proposed new ECM algorithms can be seen as a nice example of applications of those methods in the time series context, when the joint distribution of the observations is multivariate normal.

We propose two ECM type algorithms based on the conditional maximum likelihood function in our model. We propose to extend Rubin's approach for an $AR(p)$ process to our model conditionally on λ. Assuming that λ is known, conditional on the first p observations we can express the maximum likelihood estimates of ϕ and σ^2 as functions of a vector of sufficient statistics S. Taking advantage of the $AR(p)$ representation of $x_t = y_t/c_t$, the EM inference can be based on a modification of the sufficient statistics $S^{(t)}$ (see Little and Rubin 2002). Expressing estimates of $\phi(S)$ and $\sigma^2(S)$ as functions of those statistics and using modified sufficient statistics we obtain conditional EM estimates $\hat{\phi}_{EM}^{(t+1)}$ and $\hat{\sigma}_{EM}^{2\,(t+1)}$.

ECM-type algorithm

Start with $\theta^{(t)} = (\lambda^{(t)}, \phi^{(t)}, \sigma^{2(t)})$

E-step Calculate $Q(\theta|\theta^{(t)}) = E_{\theta^{(t)}}(l(y, \lambda, \phi, \sigma^2)|y_{obs})$ with possible approximation of $Q(\theta|\theta^{(t)}) \approx l(\hat{y}, \lambda, \phi, \sigma^2)$ where

$$\hat{y}_t^{(i)} = \begin{cases} y_t, & \text{if } y_t \text{ is present} \\ E\left(y_t|y_{obs}; \theta^{(i)}\right), & \text{if } y_t \text{ is missing.} \end{cases} \tag{12}$$

CM-steps

1. Obtain estimate of ϕ that is $\hat{\phi}_{EM}^{(t+1)}$.
2. Use $\hat{\phi}_{EM}^{(t+1)}$ to obtain $\hat{\sigma}_{EM}^{2\,(t+1)}$.
3. Use $\hat{\phi}_{EM}^{(t+1)}$ and $\hat{\sigma}^2_{EM}{}^{(t+1)}$ to obtain estimate of λ as

$$\hat{\lambda}_{EM}^{(t+1)} = \arg\max_{\lambda \in \Lambda} Q\left(\lambda|\lambda^{(t)}, \hat{\phi}_{EM}^{(t+1)}, \hat{\sigma}^2_{EM}{}^{(t+1)}\right).$$

Repeat the steps above until convergence is obtained.

We illustrate this algorithm with the following example.

Example 2. Let us consider the special case of model (1) when $\{x_t\}$ is an $AR(1)$ process. Let $\theta^{(t)} = (\lambda^{(t)}, \phi^{(t)}, \sigma^{2(t)})$ be a vector of starting values.

In the E-step we calculate $Q(\theta|\theta^{(t)}) = E_{\theta^{(t)}}(l(y, \lambda, \phi, \sigma^2)|y_{obs})$. To calculate $Q(\theta|\theta^{(t)})$ one may use a method proposed by Brockwell and Davis (2002). In this method, we take advantage of the fact that joint distribution of $(y_2, \ldots, y_T)$ is a multivariate normal distribution. Therefore, it is possible to approximate $Q(\theta|\theta^{(t)})$ using $l(\hat{y}, \lambda, \phi, \sigma^2)$ where

$$\hat{y}_t^{(i)} = \begin{cases} y_t, & \text{if } y_t \text{ is present} \\ E\left(y_t|y_{obs}; \theta^{(i)}\right), & \text{if } y_t \text{ is missing.} \end{cases} \tag{13}$$

Next we look at the conditional maximization steps of our algorithm. It can be seen assuming that λ is known, conditional maximum likelihood estimates for σ^2

and ϕ can be expressed using the following complete-data sufficient statistics $S = (s_3, s_4, s_5)$

$$s_3 = \sum_{t=2}^{T} \frac{y_t^2}{c_t^2}, \quad s_4 = \sum_{t=2}^{T} \frac{y_{t-1}^2}{c_{t-1}^2}, \quad s_5 = \sum_{t=2}^{T} \frac{y_{t-1}y_t}{c_{t-1}c_t}.$$

Conditional maximum likelihood estimates from a previous example given λ can be expressed as

$$\hat{\phi}_{ML} = s_5 s_4^{-1},$$

and

$$\hat{\sigma}_{ML}^2 = \left[s_3 - \hat{\phi}_{ML}^2 s_4 \right] / (T - 1).$$

Two of the conditional maximization (CM) steps will be performed to get conditional maximum likelihood estimates of (ϕ, σ^2) given λ when there are missing data. Let $\left(\phi^{(t)}, \sigma^{2(t)} \right)$ be estimates of $\left(\phi, \sigma^2 \right)$ at iteration t. This step of the algorithm calculates $\left(\phi^{(t+1)}, \sigma^{2(t+1)} \right)$ from the complete-data sufficient statistics S replaced by estimates $S^{(t)}$ from the E step.

It is worth mentioning that for a given λ the E step computes $S^{(t)} = \left(s_3^{(t)}, s_4^{(t)}, s_5^{(t)} \right)$, where

$$s_3^{(t)} = \sum_{t=2}^{T} \left[\left(\frac{\hat{y}_t^{(t)}}{c_t} \right)^2 + c_{ii}^{(t)} \right], \quad s_4^{(t)} = \sum_{t=2}^{T} \left[\left(\frac{\hat{y}_{t-1}^{(t)}}{c_{t-1}} \right)^2 + c_{i-1,i-1}^{(t)} \right],$$

$$s_5^{(t)} = \sum_{t=2}^{T} \left[\frac{\hat{y}_{t-1}^{(t)}}{c_{t-1}} \frac{\hat{y}_t^{(t)}}{c_t} + c_{i-1,i}^{(t)} \right], \tag{14}$$

$$\hat{y}_t^{(t)} = \begin{cases} y_t, & \text{if } y_t \text{ is present,} \\ E\left(y_t \,|\, y_{obs}; \theta^{(t)} \right), & \text{if } y_t \text{ is missing} \end{cases} \tag{15}$$

and

$$c_{ij}^{(t)} = \begin{cases} 0, & \text{if } y_i \text{ or } y_j \text{ is present,} \\ \frac{1}{c_i c_j} Cov\left(y_i, y_j \,|\, y_{obs}; \theta^{(t)} \right), & \text{if } y_i \text{ and } y_j \text{ are missing.} \end{cases} \tag{16}$$

Using modified sufficient statistics one obtains:

$$\hat{\phi}_{EM}^{(t+1)} = s_5^{(t)} \left(s_4^{(t)} \right)^{-1},$$

and

$$\hat{\sigma}_{EM}^2(t + 1) = \left(s_3^{(t)} - \left(\hat{\phi}_{EM}^{(t+1)} \right)^2 s_4^{(t)} \right) \Big/ (T + 1).$$

Using $\theta^{(t)}$ one obtains EM estimates of ϕ, σ^2 conditional on λ.

Then substituting those conditional EM estimates of ϕ, σ^2 into the modified log-likelihood function one obtains

$$Q\left(\theta|\theta^{(t)}\right) \approx l_c\left(\hat{y}, \lambda, \hat{\phi}_{EM}^{(t+1)}, \hat{\sigma}_{EM}^{2}{}^{(t+1)}\right)$$

$$= -\log\left(|det(C_{T-1})|\right) - \frac{T-1}{2}\log\left(\sigma_{EM}^{2}{}^{(t+1)}\right)$$

$$-\sum_{t=2}^{T} \frac{1}{\sigma_{EM}^{2}{}^{(t+1)}}\left(\frac{\hat{y}_t^{(t)}}{c_t} - \hat{\phi}_{EM}^{(t+1)}\frac{\hat{y}_t^{(t)}}{c_{t-1}}\right)^2$$

Then the conditional maximization (CM) $(t+1)$th step for λ is given as

$$\hat{\lambda}_{EM}^{(t+1)} = \arg\max_{\lambda \in \Lambda} Q\left(\lambda|\lambda^{(t)}, \hat{\phi}_{EM}^{(t+1)}, \hat{\sigma}_{EM}^{2}{}^{(t+1)}\right).$$

The function above has to be maximized numerically.

The procedure above should be repeated until convergence is obtained.

We also consider a different ECM-type algorithm and refer directly to the idea of the profile likelihood approach presented in the previous section.

ECM-type algorithm (second type)

Start with $\theta^{(t)} = (\lambda^{(t)}, \phi^{(t)}, \sigma^{2(t)})$

E-step   Calculate $Q(\theta|\theta^{(t)}) = E_{\theta^{(t)}}(l(y, \lambda, \phi, \sigma^2)|y_{obs})$ with possible approximation of $Q(\theta|\theta^{(t)}) \approx l(\hat{y}, \lambda, \phi, \sigma^2)$ where

$$\hat{y}_t^{(i)} = \begin{cases} y_t, & \text{if } y_t \text{ is present} \\ E\left(y_t|y_{obs}; \theta^{(i)}\right), & \text{if } y_t \text{ is missing.} \end{cases} \tag{17}$$

CM-steps

     1. Obtain the estimate of λ as

$$\hat{\lambda}_{EM}^{(t+1)} = \arg\max_{\lambda \in \Lambda} Q\left(\lambda|\lambda^{(t)}, \hat{\phi}_{EM}^{(t)}(\lambda), \hat{\sigma}_{EM}^{2}{}^{(t)}(\lambda)\right).$$

     2. Use $\hat{\lambda}_{EM}^{(t+1)}$ to obtain $\hat{\phi}_{EM}^{(t+1)}$

     3. Use $\hat{\lambda}_{EM}^{(t+1)}$ and $\hat{\phi}_{EM}^{(t+1)}$ to obtain $\hat{\sigma}^{2}{}_{EM}^{(t+1)}$

Repeat the steps above until convergence is obtained.

The difference between those two algorithms is obtained by making, in the second algorithm, the CM-step corresponding to estimates of λ the first CM-step and representing the EM-based estimates of ϕ and σ^2 as a function of λ, which has to be maximized numerically.

The ECM preserves the appealing convergence properties of the EM algorithm, such as its monotone convergence (see Meng and Rubin 1993; McLachlan and Krishnan 2008). The ECM algorithm converges typically more slowly than the EM algorithm in terms of number of iterations, but can be faster in total computer time.

The simulation study for comparison of different estimation procedures will be presented in the next section.

4 Simulation Study

We focus on the following model

$$y_t = c_t x_t,$$

where $x_t = \phi x_{t-1} + \varepsilon_t$, $\{\varepsilon_t\} \sim N(0, \sigma^2)$ and

$$c_t = \exp\left(\lambda_1 \cos\left(\frac{2\pi}{20}t\right) + \lambda_2 \sin\left(\frac{2\pi}{20}t\right)\right).$$

For the model under consideration, the values of the parameters were chosen $\phi = 0.5$, $\sigma^2 = 1$, $\lambda_1 = 0.4$ and $\lambda_2 = 0.5$.

In our study we consider a sample size of $n = 100$ observations. The amount of missing observations is 5, 10, 15 %, and finally 20 % of the total number of observations. For each case the simulation was repeated $Ns = 100$ times. Using our results from above we calculate the Mean Squared Error (MSE) using the following formula

$$MSE\left(\hat{\theta}\right) = E\left(\left\|\hat{\theta} - \theta\right\|^2\right),$$

where $\|x\|$ is the L_2 norm of the vector x.

The average number of iterations to reach the convergence was calculated to give some idea of the speed of convergence of the algorithms under consideration.

We have also calculated the average distance between EM-type or ECM-type algorithm based estimates and the conditional maximum likelihood estimates for the complete data.

The simulation scheme was the following. Firstly, we simulate n observations from our model and estimate the unknown vector of parameters θ using the conditional maximum likelihood approach for the complete sample. Then we choose $nm\%$ of observations to be missing. For each method we have the same observations to be missing and the same starting points being the true parameters. For the incomplete data case, we estimate unknown parameters using missing data techniques like EM and ECM algorithm.

The results of the simulations study are presented in the Tables 1, 2, 3 and 4 below:

Table 1 Results of conditional ML, EM algorithm-based and ECM algorithm-based estimation, $n = 100$, 5 % of missing observations

Method	MSE($\hat{\theta}$)	MSE($\hat{\lambda}$)	MSE($\hat{\phi}, \hat{\sigma}^2$)	Average distance	Average number of iterations
Conditional ML	0.4676363	0.02699937	0.27355		
EM	0.4717652	0.03034568	0.264452	0.009119588	13.21
ECM	0.4771919	0.03054908	0.2680076	0.01148294	11.88
ECM2	0.4717618	0.03034477	0.2644503	0.00911772	9.84

Table 2 Results of conditional ML, EM algorithm-based and ECM algorithm-based estimation, $n = 100$, 10 % of missing observations

Method	MSE($\hat{\theta}$)	MSE($\hat{\lambda}$)	MSE($\hat{\phi}, \hat{\sigma}^2$)	Average distance	Average number of iterations
Conditional ML	0.4806782	0.0281276	0.2712334		
EM	0.4689554	0.03338665	0.2485643	0.01955335	15.85
ECM	0.5315	0.03314359	0.2996184	0.01653104	13.07
ECM2	0.4689547	0.03338656	0.2485639	0.01955304	11.84

Table 3 Results of conditional ML, EM algorithm-based and ECM algorithm-based estimation, $n = 100$, 15 % of missing observations

Method	MSE($\hat{\theta}$)	MSE($\hat{\lambda}$)	MSE($\hat{\phi}, \hat{\sigma}^2$)	Average distance	Average number of iterations
Conditional ML	0.4674685	0.02830395	0.2706489		
EM	0.5349602	0.03903968	0.2973513	0.06677665	31.03
ECM	0.5944072	0.03857192	0.3513506	0.04578313	16.39
ECM2	0.531812	0.03901023	0.2942177	0.06298165	18.93

Table 4 Results of conditional ML, EM algorithm-based and ECM algorithm-based estimation, $n = 100$, 20 % of missing observations

Method	MSE($\hat{\theta}$)	MSE($\hat{\lambda}$)	MSE($\hat{\phi}, \hat{\sigma}^2$)	Average distance	Average number of iterations
Conditional ML	0.4470226	0.02800964	0.2644081		
EM	0.9342054	0.03711569	0.6753449	0.369395	103.97
ECM	0.7046426	0.03668786	0.4434781	0.1047571	20.56
ECM2	0.9529035	0.03717287	0.6939518	0.3871174	45.8

There are several comments to be made. Firstly, the most unexpected result is the smallest MSE for $\hat{\theta}$ is obtained using EM and ECM (second type) algorithms, even smaller than the MSE for $\hat{\theta}_{CML}$. This result is an extension of results obtained for the case of autoregressive processes by Noomene (2007).

The classical ECM algorithm does not perform better than the EM algorithm in terms of MSE. However we refer directly to the idea of profile likelihood and extend it to ECM-type algorithms (second type) which leads to a slightly lower MSE than in the EM algorithm case. Therefore, in our model this method gives better results than the EM algorithm. Moreover, both proposed new algorithms are more stable and converge faster than the classical EM algorithm.

Let us summarized the obtained results. Estimating the parameters of the model under consideration should be sometimes (when there is small amount of missing observations in the data) made using EM or ECM (second) algorithms. The proposed ECM (second type) algorithms outperforms existing in the literature on cyclostationary time series EM algorithm in each situation. The main advantage of the new method is in terms of lower MSE, higher stability and faster convergence than the classical EM algorithm.

Acknowledgments The chapter was written while Oskar Knapik was visiting the Department of Statistics at University of California at Davis, USA. Its kind hospitality is greatly appreciated. Oskar Knapik gratefully acknowledges the Kosciuszko Foundation for financial support for this research.

5 Appendix

Proof of Proposition 2.1

The starting point is the observation that

$$f_y(y) = |\det C|^{-1} f_x\left(C^{-1}y\right).$$

Using that fact and properties of the process (see [8]) the joint density for the complete data set can be written as

$$f_{y_t|y_{t-1},\ldots,y_{t-p}}(y;\theta) = \left|\det C_{T-p}\right|^{-1} \cdot \prod_{t=p+1}^{T} f_{x_t|x_{t-1},\ldots,x_{t-p}}\left(\frac{y_t}{c_t}\left|\frac{y_{t-1}}{c_{t-1}},\ldots,\frac{y_{t-p}}{c_{t-p}}\right.\right)$$

taking the logarithm we have

$$l_c(\theta) = \log f_{y_T,\ldots,y_{p+1}|y_p,\ldots,y_1}\left(y_T,\ldots,y_{p+1}\,\big|\,y_p,\ldots,y_1\,;\theta\right)$$

$$= -\log\left(\left|det(C_{T-p})\right|\right) + \log f_{x_T,\ldots,x_{p+1}|x_p,\ldots,x_1}\left(\frac{y_T}{c_T},\ldots,\frac{y_{p+1}}{c_{p+1}}\,\bigg|\,\frac{y_p}{c_p},\ldots,\frac{y_1}{c_1}\,;\theta\right)$$

$$= -\log\left(\left|det(C_{T-p})\right|\right) - \frac{T-p}{2}\log(2\pi) - \frac{T-p}{2}\log\left(\sigma^2\right) +$$

$$- \sum_{t=p+1}^{T} \frac{1}{\sigma^2}\left(\frac{y_t}{c_t} - \sum_{i=1}^{p}\phi_i\frac{y_{t-i}}{c_{t-i}}\right)^2 \tag{18}$$

Criterion l_θ is quadratic w.r.t. ϕ. Thus, its minimization over ϕ can be carried out analytically. The ML estimate of ϕ is, for given λ and σ^2, obtained as

$$\hat{\phi}_{ML} = \left[(Y./C)^T (Y./C)\right]^{-1}(Y./C)^T (y./c), \tag{19}$$

where $c = \left(c_{p+1},\ldots,c_T\right)^T$, $y = \left(y_{p+1},\ldots,y_T\right)^T$, Y and C are $(T-p)\times p$ Toeplitz matrices

$$Y = \begin{pmatrix} y_p & \cdots y_1 \\ \vdots & \ddots \vdots \\ y_{T-1} & \cdots y_{T-p} \end{pmatrix}; C = \begin{pmatrix} c_p & \cdots c_1 \\ \vdots & \ddots \vdots \\ c_{T-1} & \cdots c_{T-p} \end{pmatrix}. \tag{20}$$

Setting the partial derivative of $l_c(\theta)$ w.r.t. σ^2 to 0 yields, for given λ and ϕ, the ML estimate of σ^2

$$\hat{\sigma}^2_{ML} = \frac{1}{T-p}\sum_{t=p+1}^{T}\left(\frac{y_t}{c_t} - \sum_{i=1}^{p}\phi_i\frac{y_{t-i}}{c_{t-i}}\right)^2. \tag{21}$$

The likelihood function as a function of λ given ϕ and σ^2 is highly nonlinear. Therefore, an analytical solution can not be obtained. The maximization should be performed numerically. The form of the objective function comes from the work of Ghogho and Garel (1999).

References

Brockwell PJ, Davis RA (2002) Introduction to time series and forecasting. Springer, New York

Dempster AP, Laird NM, Rubin DB (1977) Maximum likelihood from incomplete data via the EM algorithm (with discussion). J Roy Statist Soc B 39:1–38

Drake C, Knapik O, Leśkow L (2013) Missing data analysis in cyclostationary model, Technical Transactions (accepted)

Gardner WA, Napolitano A, Paura L (2006) Cyclostationarity: half a century of research. Sig Process 86(4):639–697

Ghogho M, Garel B (1999) Maximum likelihood estimation of amplitude-modulated time series. Sig Process 75:99–116

Hamilton JD (1994) Time series analysis. Princeton University Press, New Jersey

Little RJA, Rubin DB (2002) Statistical analysis with missing data, 2nd edn. Wiley, New York

McLachlan GJ, Krishnan T (2008) The EM algorithm and extensions. 2nd edn. Wiley, New York

Meng X-L, Rubin DB (1993) Maximum likelihood estimation via the ECM algorithm: a general framework. Biometrika 80(2):267–278

Niu X (1996) Seasonal time series models with periodic variances. Technical report. http://stat.fsu.edu/techreports/M911.pdf

Noomene R (2007) Procedures of Parameters' estimation of AR(1) models into lineal state-space models. Proceedings of the world congress on engineering 2007 Vol II WCE 2007, London, 2–4 July 2007. http://www.iaeng.org/publication/WCE2007/WCE2007_pp995-999.pdf

Pawitan Y (2001) In all likelihood: statistical modelling and inference using likelihood. Oxford University Press, Oxford

Schaffer JL (1997) Analysis of incomplete multivariate data. Chapman & Hall/CRC

Stefanakos ChN, Athanassoulis GA (2001) A unified methodology for the analysis, completion and simulation of nonstationary time series with missing values, with application to wave data. Appl Ocean Res 23(4):207–220

Subsampling for Weakly Dependent
and Periodically Correlated Sequences

Elżbieta Gajecka-Mirek

Abstract In 1999 a new type of dependence in time series—weak dependence (see Dedecker et al. 2008) was introduced. This gives tools for the analysis of statistical procedures with very general data generating processes. One of such statistical procedures is subsampling. It can be used if statistical inference for dependent data based on asymptotic distributions is complicated or it fails. For independent data and stationary time series subsampling procedures are well investigated. Our research is focused on non-stationary—periodically correlated time series. In this chapter the subsampling consistency of self-normalized statistics for the generalization of the model introduced by McElroy and Politis (2007) is expanded.

1 Introduction

Many real life phenomena are characterized by a seasonal behavior. Popular methods that can be used to describe seasonal behavior are periodically and almost periodically correlated processes considered in Hurd (1991). On the other hand in many applications of time series analysis one is confronted with heavy tailed behavior. Such behavior is characterized by large probabilities of extremal events. Finally, contemporary time series modeling and statistical inference should take account of long range dependence. Therefore, in this chapter one will simultaneously be dealing with three features of time series: periodic non-stationarity, heavy tails and long memory. In such a framework, resampling procedures will be used to estimate the mean function Politis et al. (1999). The main goal is to show consistency of subsampling procedure in estimating a mean function for time series that are: periodically correlated, with heavy tails and long memory. From this perspective this work can considered as an extension of the results Jach et al. (2012).

E. Gajecka-Mirek (✉)
Department of Economics, State Higher Vocational School, Nowy Sacz, Poland
e-mail: egajecka-mirek@pwsz-ns.edu.pl

F. Chaari et al. (eds.), *Cyclostationarity: Theory and Methods*,
Lecture Notes in Mechanical Engineering, DOI: 10.1007/978-3-319-04187-2_4,

The starting point of this chapter is the model considered in McElroy and Politis (2007) and defined as:

$$Z_t = v_t N_t + \eta, \tag{1}$$

where v_t and N_t are independent, v_t is the i.i.d. sequence of α−stable random variable ($\alpha \in (1, 2)$) and N_t is the stationary Gaussian mean zero process with the long memory. Observe that in the model (1) $EZ_t = \eta$ because v_t and N_t are independent and N_t is mean zero. η denotes the constant that to be estimated. Moreover $E(\sigma_t) \neq 0$. It is easy to see that the model (1) is stationary.

The long memory property can be characterized as

Definition 1. Brockwell and Davis (1991) The process $\{X_t\}_{t \in Z}$ is said to have a long memory if its autocovariance function γ satisfies

$$\sum_{0 < |h| < n} \gamma(h) \sim Cn^\beta, \quad and \quad \sum_{0 < |h| < n} |\gamma(h)| = O(n^\beta)$$

where γ is autocovariance function, $\beta \in [0, 1)$, and $C > 0$.

We mark this condition: $LM(\beta)$.

Additionally it is assumed that the Gaussian mean zero process N_t is purely nondeterministic with the finite variance.

The α−stability is defined as follow:

Definition 2. Samorodnitsky and Taqqu (1994) Random variable X has **a stable** distribution if there exist parameters: $0 < \alpha \leq 2$, $-1 \leq \beta \leq 1$, $\sigma > 0$ and $\mu \in R$ such that characteristic function $\varphi(\cdot)$ of X has the form:

$$\varphi(t) = \begin{cases} \exp\{-\tau^\alpha |t|^\alpha (1 - i\beta\, sgn(t)\frac{\pi\alpha}{2}) + i\mu t\}, & \alpha \neq 1 \\ \exp\{-\tau |t|(1 + i\beta\frac{2}{\pi}\, sgn(t) \ln |t|) + i\mu t\}, & \alpha = 1 \end{cases}$$

α is the stability index, β is the skewness parameter, τ is the scale parameter and μ is the location parameter.

From Guégan (2001) it is known that in the Gaussian long memory process the special type of time series dependence need to be considered—the weak dependence.

Definition 3. Dedecker et al. (2008) Let $(E, \| \cdot \|)$ be a normed space. We assume that a function $h : E^w \longrightarrow R$ belongs to the class $\mathscr{L} = \{h : E^w \to R, \| h \|_\infty \leq 1, Lip(h) < \infty\}$, where $Lip(h) = \sup_{x \neq y} \frac{|h(x) - h(y)|}{\|x - y\|_1}$ and $\| x \|_1 = \sum_{i=1}^w \| x_i \|$, $w \in N$.

A sequence $\{X_n\}_{n \in N}$ of random variables taking values in E^w is called $(\varepsilon, \mathscr{L}, \Psi)-$ **weakly (ε−weakly) dependent** if there exists $\Psi : \mathscr{L} \times \mathscr{L} \times N^* \times N^* \to R$ ($N^* = N \cup \{0\}$) and a sequence $\{\varepsilon_r\}_{r \in N}$ ($\varepsilon_r \to 0$) such that for any $(f, g) \in \mathscr{L} \times \mathscr{L}$, and $(u, v, r) \in N^{*2} \times N$

$$|Cov(f(X_{i_1}, ..., X_{i_u}), g(X_{j_1}, ..., X_{j_v}))| \leq \Psi(f, g, u, v)\varepsilon_r$$

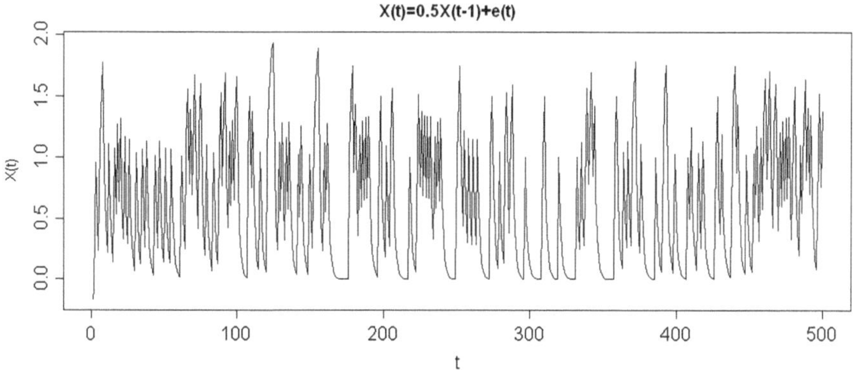

Fig. 1 AR(1) process with Bernoulli innovations with parameter $s = 0.3$

whenever $i_1 < i_2 < \cdots < i_u \le r + i_u \le j_1 < j_2 < \cdots < j_v$.

The following distinct functions Ψ yield θ weak dependence coefficients:
if $\Psi(f, g, u, v) = vLip(g)$ then $\varepsilon(r) = \theta(r)$.

The notion of weak dependence gives us tools for the analysis of statistical procedures with very general processes (e.g.: Bernoulli shifts or Markov processes driven by discrete innovations).

In what follows, the model (1) is extended into the time series with periodically correlated autocovariance, according to the definition below.

Definition 4. Time series $\{X_t\}_{t \in Z}$ is Periodically Correlated (PC) if the mean $\mu_X(t) = E(X_t) = \mu_X(t + T)$ and the autocovariance function $B_X(t, h) = cov(X_t, X_{t+h}) = B_X(t + T, h)$, for all $h \in Z$.

Example 1 *Bernoulli shift:* $X_n = H(\xi_n, \xi_{n-1}, \ldots)$ *(with* $H(x) = \sum_{k=0}^{\infty} 2^{-(k+1)} x_k$*) is not mixing but is weakly dependent.*

$X_n = \sum_{k=0}^{\infty} 2^{-(k+1)} \xi_{n-k}$, *where* ξ_{n-k} *is the kth digit in the binary representation of the uniformly chosen number* $X_n = 0. \xi_n \xi_{n-1} \ldots \in [0, 1]$*. Such* X_n *is deterministic function of* X_0*, so the event* $A = (X_0 \le \frac{1}{2})$ *belongs to the* $\sigma -$ *algebras:* $\sigma(X_t, t \le 0)$ *and* $\sigma(X_t, t \ge n)$*. From definition:* $\alpha(n) \ge | P(A \cap A) - P(A)P(A) | = \frac{1}{2} - \frac{1}{4} = \frac{1}{4}$*. The* $\{X_n\}$ *is weakly dependent, because:*

Lemma 1 *Dedecker et al. (2008) Bernoulli shifts are* $\theta -$*weakly dependent with* $\theta(r) \le 2\delta_{[r/2]}$*, where* $\{\delta_r\}_{r \in N}$ *is defined by:* $E \mid H(\xi_{t-j}, j \in Z) - H(\xi_{t-j} 1_{|j| \le r}, j \in Z) \mid$*.*

Example 2 *Example of the model which satisfied weak dependence definition (but not fulfilled mixing conditions) is AR(1) model defined as:* $X_t = aX_{t-1} + \varepsilon_t$*, where* $|a| < 1$ *and innovations* $\{\varepsilon_t\}$ *are i.i.d. Bernoulli variables with parameter* $s = P(\varepsilon_t = 1) = 1 - P(\varepsilon_t = 0)$ *(illustration—Fig. 1).*

The extension of the model (1) investigated in this chapter is:

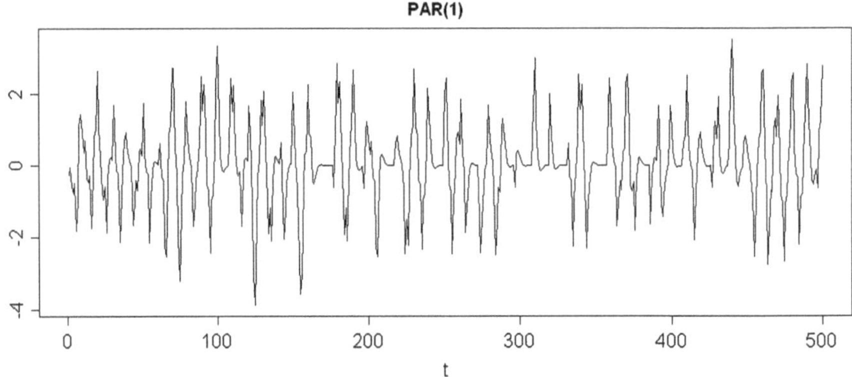

Fig. 2 PAR(1)—AR(1) process multiplied by periodical function $f(t) = 2cos(2\pi t/10)$

$$X_t = \sigma_t G_t + \eta_t,$$

where σ_t and G_t are independent, σ_t, is i.i.d. mean μ different from zero and has the marginal distribution of an $\alpha-$stable random variable. Moreover G_t is periodically correlated time series with known period T and can be written as $G_t = f_t N_t$ for a long memory, stationary mean zero Gaussian process N_t, and f_t- bounded and scalar periodic sequence $f_t = f_{t+T}$, $EX_t = \eta_t (= \eta(t))$ is deterministic and periodic function with the same, known period T as f_t (Fig. 2).

It is easy to see that above model is periodically correlated. It is shown in Sect. 2.

One of the statistical methods to analyze a time series are resampling methods. The resampling procedures for independent data and stationary time series are well developed. But there is still not enough of research in field of nonstationary models, although some results being published (e.g. Dudek 2011; Lenart 2008).

For the model (2) the subsampling (one of the resampling methods) is considered to approximate an asymptotic distribution of a self-normalized sample mean's vector.

The idea of subsampling is as follows: the statistic $\vartheta_n(\hat{\varphi}_n - \varphi)$ is recomputed over "short", overlapping blocks of length b (b depends on n, where n is the length of the sample, $\hat{\varphi}_n$ is the estimator of unknown parameter φ).

$n - b + 1$ statistics $\vartheta_b(\hat{\varphi}_{n,b,t} - \hat{\varphi}_n)$ are obtained where $\hat{\varphi}_{n,b,t}$ are subsampling versions of the estimator $\hat{\varphi}_n$.

Then empirical distributions $L_{n,b}(x) = \frac{1}{n-b+1} \sum_{t=1}^{n-b+1} 1_{\{\vartheta_b(\hat{\varphi}_{n,b,t}-\hat{\varphi}_n)\leq x\}}$ are used to approximate the asymptotic distribution $J(x)$ of the estimator $\vartheta_n(\hat{\varphi}_n - \varphi)$.

2 The Model

Let the time series $\{X_t\}_{t\in Z}$ be defined as:

$$X_t = \sigma_t G_t + \eta_t, \tag{2}$$

where

1. σ_t and G_t are independent,
2. Let $\sigma_t = \sqrt{\varepsilon_t}$, where ε_t are i.i.d. and have the marginal distribution of an $\alpha/2 - stable$ random variable with skewness parameter 1, scale parameter $\tau = (cos(\pi\alpha/4))^{2/\alpha}$, and location parameter zero,
3. $E(\sigma_t) = \mu$ different from zero,
4. G_t is periodically correlated time series with period T and can be written as $G_t = f_t N_t$ for a long memory, purely non-deterministic, stationary mean zero Gaussian process N_t, and $f_t -$ bounded and scalar periodic sequence $f_t = f_{t+T}$,
5. G_t is a long memory with parameter $\beta \in (0, 1)$,
6. $EX_t = \eta_t (= \eta(t))$ is deterministic and periodic with the same period T as the function f_t.

Proposition 1 The model (2) is periodically correlated.

Proof

$$B_X(t + T, h) = f_{t+T} f_{t+T+h} B_Z(t + T, h)$$
$$= f_{t+T} f_{t+T+h} B_Z(h) = f_t f_{t+h} B_Z(t, h) = B_X(t, h),$$

$B_X(t, h)$ and $B_Z(t, h)$ are the autocovariance functions of respectively the model (2) and the model (1). $\qquad\square$

X_t has symmetric and $\alpha-$stable marginal distribution with the scale parameter $f_t\sqrt{\gamma_N(0)/2}$, (γ_N is autocovariance function of N_t from definition of model (2)) for $t = 1, 2, ..., NT$ and the mean of σ_t exists and is equal to μ (see Samorodnitsky and Taqqu 1994).

It is obvious, from the definition of the model (2), that autocovariance function $Cov[X_t, X_{t+h}] = B_X(t, h) = B_X(t + T, h)$ is infinite for $h = 0$ but for $h \neq 0$, $Cov[X_t, X_{t+h}] = \mu^2 E G_t G_{t+h} = f_t f_{t+h} B_Z(h) = \mu^2 f_t f_{t+h} B_N(h) = \mu^2 f_t f_{t+h} \gamma_N(h)$.

3 Central Limit Theorem

To apply subsampling method we must know: if a non-degenerate asymptotic distribution of the statistic exists and if subsampling method is consistent. In this section we will focus on the first of above conditions and introduce generalization of CLT of McElroy and Politis (2007).

Consider $\bar{X}_N(s)$ as an estimator of $\eta_s = \eta_N(s)$:

$$\hat{\eta}_N(s) = \bar{X}_N(s) = \frac{1}{N} \sum_{p=0}^{N-1} X_{s+pT}, \quad s = 1, 2, ..., T,$$

where T is the period, constant.

Let: $N(\bar{X}_N(s) - \eta_N(s)) = \sum_{p=0}^{N-1} Y_{s+pT}, \quad s = 1, 2, ..., T$ where $Y_t = X_t - \eta(t) = \sigma_t G_t$.

Define

$$\zeta = max\{1/\alpha, (\beta + 1)/2\}.$$

α is a tail parameter and β is a long memory parameter.

In this chapter it is assumed that the number of observation is NT. From the definition of the model (2) $\sum_{0<|h|<NT} |\gamma_N(h)| = O((NT)^\beta)$. It follows that

$\sum_{0<|h|<N} |\gamma_N(hT)| \leq \sum_{0<|h|<NT} |\gamma_N(h)| = O((NT)^\beta) = O(N^\beta)$, since T is constant.

Theorem 1. *Suppose the assumptions 1.–6. from Sect. 2 hold, and assume that the long memory condition $LM(\beta)$ holds for $\beta \in [0, 1)$ with the same β for each $s = 1, 2, ..., T$. Then*

$$N^{-\zeta} \sum_{p=0}^{N-1} Y_{s+pT} \xrightarrow{d} \begin{cases} S(s) & if \ 1/\alpha > (\beta + 1)/2 \\ V(s) & if \ 1/\alpha < (\beta + 1)/2 \\ S(s) + V(s) & if \ 1/\alpha = (\beta + 1)/2. \end{cases} \tag{3}$$

$\xrightarrow{d}$ denotes convergence in distributions. $S(s)$ is a SαS variable with zero location parameter, and scale $f_s\sqrt{\gamma_N(0)/2}, \ s = 1, .., T$. $V(s)$ is a mean zero Gaussian variable with variance $\tilde{C}(s)\mu^2/(\beta + 1)$. Moreover $S(s)$ and $V(s)$ are independent.

Sketch of proof in Appendix

First theorem concerns convergence of the sample means. Observe that if we extend Theorem 1 and take $\alpha = 2$ and $\beta = 0$ we obtain the classical central limit theorem.

In the second theorem joint asymptotic behavior for normalized sample means and sample variances is introduced.

Theorem 2. *Suppose the assumptions 1.–6. from Sect. 2 hold, and assume that the long memory condition $LM(\beta)$ holds for $\beta \in [0, 1)$ with the same β for each $s = 1, 2, ..., T$. Then the sample first and second moments of the $\{Y_t\}$ series, normalized by N^ζ, converge jointly to absolutely continuous random variables.*

$$\left(N^{-\zeta} \sum_{p=0}^{N-1} (X_{s+pT} - \eta_N(t)), \ N^{-2\zeta} \sum_{p=0}^{N-1} (X_{s+pT} - \bar{X}_N(s))^2\right)$$

$$\xrightarrow{d} \begin{cases} (S(s), U(s)) & if \ 1/\alpha > (\beta + 1)/2 \\ (V(s), 0) & if \ 1/\alpha < (\beta + 1)/2 \\ (S(s) + V(s), U(s)) & if \ 1/\alpha = (\beta + 1)/2. \end{cases} \tag{4}$$

$S(s)$, $V(s)$ are defined as in Theorem 1. $U(s)$ is $\alpha/2$ stable with zero location parameter, skewness one, and scale proportional to $\tau f_s^2 \gamma_N(0)$, for $s = 1, 2, ..., T$. $V(s)$ is independent of $U(s)$, but $S(s)$ and $U(s)$ are dependent.

Sketch of proof in Appendix

Observe that in the case $1/\alpha > (\beta + 1)/2$ the sample variances can be used to studentize the $\bar{X}_N(s)$. While $1/\alpha < (\beta + 1)/2$ another normalization needs to be found.

Let us define

$$\widehat{LM}(\rho, s) = \Big| \sum_{|h|=1}^{[N^\rho]} \frac{1}{N - |h|} \sum_{p=0}^{N-|h|} (X_{s+pT} X_{s+pT+hT} - \bar{X}_N^2(s)) \Big|^{1/\rho},$$

where $\rho \in (0, 1)$ $s = 1, 2, ..., T$.

As in the chapter McElroy and Politis (2007) we will use $\widehat{LM}(\rho, t)$ as an additional normalization of our sample means.

Next theorem investigates asymptotic behavior of $\widehat{LM}(\rho, t)$.

Theorem 3. *Suppose the assumptions 1.–6. from Sect. 2 hold, and assume that the long memory condition $LM(\beta)$ holds for $\beta \in [0, 1)$ with the same β for each $s = 1, 2, ..., T$. Let $\rho \in (0, 1)$. Then $\widehat{LM}(\rho, s)$ converges in probability to a constant at rate N^β. In particular,*

$$N^{-\beta} \widehat{LM}(\rho, s) \xrightarrow{P} \mu^{2/\rho} C(s)^{1/\rho}.$$

Proof in Appendix

Main result—Theorem 4 is conclusion of Theorems 1, 2 and 3.

Theorem 4. **Central Limit Theorem**
Suppose the assumptions 1.–6. from Sect. 2 hold, and assume that the long memory condition $LM(\beta)$ holds for $\beta \in [0, 1)$ with the same β for each $s = 1, 2, ..., T$. Let $\rho \in (0, 1)$. Then the following joint weak convergence holds:

$$\Big(N^{-\zeta} \sum_{p=0}^{N-1} (X_{s+pT} - \eta_N(s)), \ N^{-2\zeta} \sum_{p=0}^{N-1} (X_{s+pT} - \bar{X}_N(s))^2, \ N^{-2\zeta + \Gamma} \widehat{LM}(\rho, s) \Big)$$

$$\xrightarrow{d} \begin{cases} (S(s), U(s), 0), & \text{if } 1/\alpha > (\beta + 1)/2 \\ (V(s), 0, \mu^{2/\rho} C(s)^{1/\rho}), & \text{if } 1/\alpha < (\beta + 1)/2 \\ (S(s) + V(s), U(s), \mu^{2/\rho} C(s)^{1/\rho}), & \text{if } 1/\alpha = (\beta + 1)/2. \end{cases} \tag{5}$$

The normalized statistic also converges weakly. Define $P_N(s)$ by

$$P_N(s) = \frac{\sqrt{N}(\bar{X}_N(s) - \eta_N(s))}{\sqrt{\frac{1}{N}\sum_{p=0}^{N-1}(X_{s+pT} - \bar{X}_N(s))^2 + \widehat{LM}(\rho, s)}}$$

and an absolutely continuous random variable $Q(s)$ by

$$Q(s) = \begin{cases} S(s)/\sqrt{U(s)}, & \text{if } 1/\alpha > (\beta+1)/2 \\ V(s)/\mu^{1/\rho}C(s)^{1/2\rho}, & \text{if } 1/\alpha < (\beta+1)/2 \\ (S(s)+V(s))/\sqrt{U(s) + \mu^{2/\rho}C(s)^{1/\rho}}, & \text{if } 1/\alpha = (\beta+1)/2. \end{cases}$$

Then

$$P_N(s) \xrightarrow{d} Q(s). \tag{6}$$

Proof in Appendix

4 Consistency of the Subsampling

From the Central Limit Theorem we know that a non-degenerate asymptotic distribution of the normalized statistic exists. Provided the subsampling estimator is consistent, the subsampling method can be used to estimate the vector of the means.

Assume that for the model (2) the following seven conditions hold:

1. The statistic $\widehat{\eta}_N(s)$ is an estimator of unknown parameter $\eta_N(s)$, with studentization $P_N(s) = \frac{\sqrt{N}(\widehat{\eta}_N(s) - \eta_N(s))}{\widehat{\sigma}_N(s)}$, $s = 1, 2, ..., T$, where
$\widehat{\sigma}_N(s) = \sqrt{\frac{1}{N}\sum_{p=0}^{N-1}(X_{s+pT} - \widehat{\eta}_N(s))^2 + \widehat{LM}(\rho, s)}$.
2. The statistics $P_N(s)$, $s = 1, 2, ..., T$ converge weakly to the limit random variables with the cumulative distribution functions $L(s)$.
3. The empirical distribution functions $L_N(s)$, are computed from the subsamples $\{X_{s+pT}\}$, $(s = 1, 2, ..., T, p = 0, ..., N-1)$ of the length N from the sample of the length NT, where $X_t = f_t Z_t$ is PC and ε—weakly dependent with the weak dependence parameters ε_r (by the Proposition 1 of Jach et al. (2012) $\{Z_t\}$ is θ—weakly dependent), Z_t is strictly stationary time series.
4. $L_N(s)(x) \to L(s)(x)$ if $N \to \infty$, $s = 1, 2, ..., T$.
5. $\sqrt{N}(\widehat{\eta}_N(s) - \eta_N(s))$ converge weakly to $Z(s)$, for all $s = 1, 2, ..., T$. $Z(s)$ are random variables and $\widehat{\eta}_N(s)$ converge weakly to $W(s)$, $s = 1, 2, ..., T$ where $W(s)$, are positive random variables with the probability 1.
6. $\frac{b}{N} \to 0$, if $b \to \infty$ and $N \to \infty$.
$b_N = b$ is the length of subsampling subseries: $X_{s+qT}, X_{s+(q+1)T}\cdots,$ $X_{s+(q+b-1)T}, q = 0, 1, ..., N - b$.
7. $\sum_{r=0}^{\infty} \varepsilon(r)^{\frac{2}{3}} < \infty$, $\lim_{n\to\infty} \frac{b}{N}(1 \vee (\frac{L(b)^4}{b})^{\frac{1}{3}} \vee (\frac{L(b)}{b})^{\frac{2}{3}}) = 0$, $\varepsilon(r)$ is the weak dependence parameter.

$a \vee b$ is the maximum of the numbers $a, b \geq 0$. $L(b) = LipZ_{N,b,q}(s)$, where

$$Z_{N,b,q}(s) = \sqrt{b}(\widehat{\eta}_{N,b,q}(s) - \eta_N(s))/\widehat{\sigma}_{N,b,q}(s)$$

$$\widehat{\eta}_{N,b,q}(s) = b^{-1} \sum_{l=s+qT}^{s+qT+b-1} X_l,$$

$$\widehat{\sigma}_{N,b,q}^2(s) = b^{-1} \sum_{l=s+qT}^{s+qT+b-1} (X_l - \widehat{\eta}_{N,b,q}(s))^2 + \widehat{LM}_{N,b,q}(\rho, s),$$

$$\widehat{LM}_{N,b,q}(\rho, s) = |\sum_{|h|=1}^{[b^\rho]} \frac{1}{b - |h|} \sum_{l=s+qT}^{s+qT+b-1-|h|} (X_l X_{l+h} - (\widehat{\eta}_{N,b,q}(s))^2)|^{1/\rho_s},$$

$\rho \in (0, 1)$.

Theorem 5. *Consistency of the subsampling estimator*
Under the above seven conditions we have:

1. *If x is the point of the continuity $L(s)(x)$, then $L_{N,b}(s)(x) \xrightarrow{p} L(s)(x)$*
2. *If $L(s)$ is continuous then $\sup_x |L_{N,b}(s)(x) - L(s)(x)| \xrightarrow{p} 0$.*
3. *If $L(s)$ is continuous in $c(1 - p)$, then if $N \to \infty$*

$$P(\sqrt{N}(\widehat{\eta}_N(s) - \eta_N(s))/\widehat{\sigma}_N(s) \leq c_{N,b}(1 - p)) \to 1 - p.$$

Proof in Appendix

5 Conclusions

In this chapter the model of the periodically correlated, long memory and heavy tailed time series was described. Estimator of the vector sample means of this particular model was introduced and Central Limit Theorem was proved. Furthermore, consistency of chosen method of estimation, the subsampling, was investigated. The question to be answered in further research is the extension of the model (2) to the class almost periodically correlated processes.

Appendix

In the appendix the proofs of some results are presented.

Proof Theorem 1. (sketch)
The proof follows as in McElroy and Politis (2007) and is almost repetition of it, hence only main points will be repeated.

Let $\mathscr{E}$ be the $\sigma-$field: $\mathscr{E} = \sigma(\varepsilon) = \sigma(\varepsilon_t, t \in Z)$. Let $\mathscr{G}$ be the $\sigma-$field: $\mathscr{G} = \sigma(G) = \sigma(G_t, t \in Z)$. From the assumption 1. in the definition of the model (2) the $\sigma-$fields $\mathscr{E}$ and $\mathscr{G}$ are independent with respect to the probability measure P.

Let assume that $\beta = 0$, it follows $LM(0)$. The characteristic function for the normalized sum is

$$Eexp\{ivN^{-1/\alpha} \sum_{p=0}^{N-1} Y_{s+pT}\} = E[E[exp\{ivN^{-1/\alpha} \sum_{p=0}^{N-1} \sigma_{s+pT} G_{s+pT}\}|\mathscr{E}]]$$

where v is any real number and $s = 1, 2, ..., T$. Let investigate the inner conditional characteristic function. From the properties of Gaussian characteristic function

$$E[exp\{ivN^{-1/\alpha} \sum_{p=0}^{N-1} \sigma_{s+pT} G_{s+pT}\}|\mathscr{E}]$$

$$= exp\{-\frac{(vN^{-1/\alpha})^2}{2} \sum_{p,q=0}^{N-1} \sigma_{s+pT}\sigma_{s+qT}f_{s+pT}f_{s+qT}\gamma_N(T(p-q))\},$$

$$s = 1, ..., T.$$

The double sum is divided into the diagonal and off-diagonal terms:

$$N^{-\frac{2}{\alpha}} \left(\sum_{p=0}^{N-1} \sigma_{s+pT}^2 \gamma_G(0) + \sum_{p \neq q} \sigma_{s+pT}\sigma_{s+qT} \gamma_G((p-q)T) \right) \tag{7}$$

The second part of (7) tends to 0 in probability (see McElroy and Politis (2007)). So off-diagonal part of (7) tends to zero in probability by the Markov inequality and hence, as it is known, in distributions. The characteristic function of the first term of (7) is the characteristic function of a $s\alpha s$ variable with scale $\sqrt{\gamma_G(0)/2} = f_s\sqrt{\gamma_N(0)/2}$, for $s = 1, 2, ..., T$ (McElroy and Politis 2007; Billingsley 1995; Samorodnitsky and Taqqu 1994).

In the case $1/\alpha > (\beta+1)/2$ the second term of (7) is $O(N^{1-2/\alpha}N^{\beta})$ which tends to zero as $N \to \infty$. The rest of the proof is identical as in the case $\beta = 0$.
In the case $1/\alpha < (\beta+1)/2$ the formula (7) becomes

$$N^{-(\beta+1)}\Big(\sum_{p=0}^{N-1}\sigma_{s+pT}^2\gamma_G(0)+\sum_{q\neq p}\sigma_{s+pT}\sigma_{s+qT}\gamma_G((p-q)T)\Big). \qquad (8)$$

The first term is $O_P(N^{2/\alpha-(\beta+1)})$ and tends to zero as $N\to\infty$.

The limiting characteristic function of the second (Lemma 1 McElroy and Politis 2007, Lemma 3.1 of Taqqu 1975) is

$$Eexp\{-\frac{v^2}{2}\frac{\tilde{C}(t)\mu^2}{\beta+1}\}=exp\{-\frac{v^2}{2}\frac{\tilde{C}(s)\mu^2}{\beta+1}\}$$

which is characteristic function of a mean zero Gaussian with variance $\tilde{C}(s)\mu^2/(\beta+1)$.

The case $1/\alpha=(\beta+1)/2$ is the combination of the two above cases. From the Slutsky's Theorem we get weak convergence the sum of two independent random variables. The characteristic function is in the form: $Eexp\{-\frac{v^2}{2}\big(\gamma_G(0)\varepsilon_\infty+\tilde{C}(t)\mu^2/(\beta+1)\big)\}=exp\{-|v|^\alpha(\gamma_G(0)/2)^{\alpha/2}\}\cdot exp\{-\frac{v^2}{2}\frac{\tilde{C}(s)\mu^2}{\beta+1}\}$, and indeed this is a characteristic function of the sum of a stable S and a normal V variables. $\qquad\square$

Proof Theorem 2. (sketch)
The proof follows as in McElroy and Politis (2007) and is almost repetition of it, hence only main points will be repeated.

$$\sum_{p=0}^{N-1}(X_{s+pT}-\bar{X}_N(s))^2=\sum_{p=0}^{N-1}(Y_{s+pT}-\bar{Y}_N)(s)^2=\sum_{p=0}^{N-1}Y_{s+pT}^2-N\bar{Y}_N^2(s),\ s=1,...,T$$

The second term is, by Theorem 2, bounded in probability of order $N^{2\zeta-1}$. Therefore

$$N^{-2/\alpha}\sum_{p=0}^{N-1}(X_{s+pT}-\bar{X}_N(s))^2=o_P(1)+N^{-2/\alpha}\sum_{p=0}^{N-1}Y_{s+pT}^2.$$

In the proof the joint Fourier/Laplace Transform of the first and second sample moments (see Fitzsimmons and McElroy 2006) is considered. For any real θ and $\phi>0$,

$$Eexp\{i\theta N^{-\zeta}\sum_{p=0}^{N-1}Y_{s+pT}-\phi N^{-2\zeta}\sum_{p=0}^{N-1}Y_{s+pT}^2\}$$

$$=E[exp\{-\frac{1}{2}N^{-\zeta}\sum_{p,q}^{N-1}\sigma_{s+pT}\sigma_{s+qT}\gamma_G((p-q)T)(\theta+\sqrt{2\phi}W_{s+pT})$$
$$(\theta+\sqrt{2\phi}W_{s+qT})\}].$$

The sequence of random variables W_s is i.i.d. standard normal, and is independent of the Y_s series. The information about W_s is denoted by $\mathcal{W}$. The double sum in the Fourier/Laplace Transform is broken into diagonal and off-diagonal terms. The off-diagonal term is

$$N^{-2\zeta} \sum_{\substack{|h|>0}}^{N-1} \sum_{p=0}^{N-|h|} \sigma_{s+pT}\sigma_{s+pT+hT}(\theta + \sqrt{2\phi}W_{s+pT})(\theta + \sqrt{2\phi}W_{s+pT+hT})\gamma_G(hT) \quad (9)$$

In the case $2/\alpha > \beta + 1$ absolute expectation of (9) is

$$E[E[N^{-2/\alpha}\Big| \sum_{\substack{|h|>0}}^{N-1} \sum_{p=0}^{N-|h|} \sigma_{s+pT}\sigma_{s+pT+hT}(\theta + \sqrt{2\phi}W_{s+pT})$$

$$\times (\theta + \sqrt{2\phi}W_{s+pT+hT})\gamma_G(hT)\Big| |\mathcal{W}]]$$

$$\leq \mu^2 (E|\theta + \sqrt{2\phi}W|)^2 N^{1-2/\alpha}K^2 \sum_{\substack{|h|>0}}^{N-1}(1-|h|/N)|\gamma_N(hT)|$$

The summation over h is $O(N^\beta)$ by the long memory assumption: $LM(\beta)$. Hence right side of above inequality is $O(N^{\beta+1-2/\alpha})$ and tends to zero as $N \to \infty$. By the Markov inequality the off-diagonal term tends to zero in probability. (K is the limitation of periodic function f_s.)

In the case $2/\alpha < \beta + 1$ (9) can be rewritten as

$$N^{-(\beta+1)} \sum_{\substack{|h|>0}}^{N-1} \sum_{p=0}^{N-|h|} (\sigma_{s+pT}\sigma_{s+pT+hT} - \mu^2)(\theta + \sqrt{2\phi}W_{s+pT})(\theta + \sqrt{2\phi}W_{s+pT+hT})$$

$$\times \gamma_G(hT) + \mu^2 N^{-(\beta+1)} \sum_{\substack{|h|>0}}^{N-1} \sum_{p=0}^{N-|h|} \gamma_G(hT)(\theta + \sqrt{2\phi}W_{s+pT})(\theta + \sqrt{2\phi}W_{s+pT+hT}).$$

The first term of (9) by the Lemma 1 McElroy and Politis (2007) and the Strong Law of Large Numbers is bounded in probability of order $N^{-(\beta+1)+\beta+\delta+1/\alpha}$ which tends to zero for small enough δ. The absolute expectation the second term of (9) is $O(N^{-(\beta+1)+\beta+1/2})$ and it tends to zero (McElroy and Politis 2007). The first term tends to constant (see the proof of Theorem 1 McElroy and Politis (2007)).

In the case $2/\alpha = \beta+1$, we see that the off-diagonal terms converge in probability to $\tilde{C}(t)\frac{\mu^2\theta^2}{\beta+1}I_{2/\alpha\leq\beta+1}$. For fixed $s = 1, 2, ..., T$ the off-diagonal terms tend to a constant.

The characteristic function of the diagonal terms is examined separately (by Dominated Convergence Theorem and above fact). Let $V_{s+pT} = \theta + \sqrt{2\phi}W_{s+pT}$

$$E[exp\{-\frac{1}{2}\gamma_G(0)N^{-2\zeta}\sum_{p=0}^{N-1}\sigma^2_{s+pT}V^2_{s+pT}\}]$$

$$= E[exp\{-(\gamma_G(0)/2)^{\alpha/2}N^{-\alpha\zeta}\sum_{p=0}^{N-1}|V_{s+pT}|^{\alpha}\}].$$

While $2/\alpha < \beta+1$ the sum $N^{-\alpha\zeta}\sum_{p=0}^{N-1}|V_{s+pT}|^{\alpha} \xrightarrow{p} 0$. While $2/\alpha \geq \beta+1$ (by the Law of Large Numbers) $N^{-\alpha\zeta}\sum_{p=0}^{N-1}|V_{s+pT}|^{\alpha} \xrightarrow{p} E|V|^{\alpha}$. By the Dominated Convergence Theorem, the limit as $N \longrightarrow \infty$ can be taken through the expectation, so that $E[exp\{-(\gamma_G(0)/2)^{\alpha/2}N^{-\alpha\zeta}\sum_{p=0}^{N-1}|V_{s+pT}|^{\alpha}\}] \to exp\{-(\gamma_G(0)/2)^{\alpha/2}E|\theta + \sqrt{2\phi}N|^{\alpha}1_{2/\alpha\geq\beta+1}\}$. Using the Fourier/Laplace Transform and argumentation as in McElroy and Politis (2007) we get the proof. $\square$

Proof Theorem 3.

The proof follows as in Jach et al. (2012) and is almost repetition of it.

Using argumentation as in McElroy and Politis (2007) and in Jach et al. (2012) we obtain for $s = 1, ..., T$

$$N^{-\beta\rho}(\widehat{LM}(\rho, s))^{\rho} = o_P(1)$$

$$+ \left|N^{-\beta\rho}\sum_{|h|>0}^{[N^{\rho}]}\frac{1}{N-|h|}\sum_{p=0}^{N-|h|}Y_{s+pT}Y_{s+pT+hT}\right| \xrightarrow{P} \mu^2 C(s).$$

$\square$

Proof Theorem 4.

The convergence of (5) follows from Theorem 2 and Theorem 3 and Slutsky Theorem. The convergence of (6) follows from the continuous mapping theorem (denominators are different from zero). $\square$

Proof Theorem 5.

Lets consider a sequence of statistics $Z_N(s) = \sqrt{N}(s)(\hat{\eta}_N(s) - \eta(s))/\hat{\sigma}_N(s)$, for fixed $s = 1, 2, ..., T$ and $N = 1, 2, ...$

$L_N(s)(x) = P(Z_N(s) \leq x)$ is cumulative distribution function of $Z_N(s)$.

From Theorem 2 we know that assumption 4 of the Theorem 5 is fulfill, hence there exist

$$r_N(s) = sup_{x\in R}|L_N(s)(x) - L(s)(x)| \longrightarrow 0, \quad N \to \infty$$

and $\| L' \|_{\infty} < \infty$.

L' denotes the density of limit distribution.

For overlapping samples the number of subsamples:

$Y_{b,q}(s) = (X_{s+qT}, X_{s+(q+1)T}..., X_{s+(q+b-1)T})$, $q = 0, 1, ..., N - b$ and the number of subsampling statistics:

$$P_{N,b,q}(s) = \sqrt{b}(\hat{\eta}_{N,b,q}(s) - \hat{\eta}_N(s))/\hat{\sigma}_{N,b,q}(s) \text{ is } N - b + 1.$$

Above statistics are used to approximate the distributions $L_N(s)(x)$ by empirical distribution functions: $L_{N,b,q}(s)(x) = \frac{1}{N-b+1} \sum_{q=0}^{N-b} I_{\{P_{N,b,q}(s) \leq x\}}$.

For $Z_{N,b,q}(s)$ let define rough subsampled statistics:

$$U_{N,b,q}(s)(x) = \frac{1}{N-b+1} \sum_{q=0}^{N-b} I_{\{Z_{N,b,q}(s) \leq x\}}.$$

From Theorem 11.3.1 Politis et al. (1999) for Heavy Tails it is known that

$$\forall x \in R \ |L_{N,b,q}(s)(x) - U_{N,b,q}(s)(x)| \xrightarrow{p} 0.$$

It follows that it is enough to investigate only the variance of $U_{N,b,q}(s)$, $s = 1, ..., T$.

By the Theorem 2 Doukhan et al. (2011), under Assumptions 4. and 7. we have:

$$lim_{N\to\infty}|E[U_{N,b,q}(s)(x) - E[U_{N,b,q}(s)(x)]]^2| = 0.$$

It implies that $Var(U_{N,b,q}(s)(x))$ tends to zero, it proves point 1. of the Theorem 5.

To prove the point 2. of the Theorem 5 we also use the Theorem 2 Doukhan et al. (2011).

$$lim_{N\to\infty}sup_{x\in R}|U_{N,b,q}(s)(x) - L(s)(x)| = 0,$$

in probability. The proof of point 3. given 1. and under assumption of model (2) is very similar to the proof of 3. in the Theorem 11.3.1 of McElroy and Politis (2007).

$\square$

References

Billingsley P (1995) Probability and measure. Wiley, New York

Brockwell P, Davis R (1991) Time series: theory and methods. Springer, New York

Dedecker J, Doukhan P, Lang G, León JR, Louhichi S, Prieur C (2008) Weak dependence: with examples and applications (Lecture notes 190 in statistics). Springer, New York

Doukhan P, Prohl S, Robert CY (2011) Subsampling weakly dependent times series and application to extremes. TEST 20(3):487–490

Dudek AE, Leśkow J (2011) A bootstrap algorithm for data from a periodic multiplicative intensity model. Comm Statist Theor Methods 40:1468–1489

Fitzsimmons P, McElroy T (2006) On joint fourier-laplace transforms. Mimeo. http://www.math.ucsd.edu/politis/PAPER/FL.pdf

Guégan D, Ladoucette S (2001) Non-mixing properties of long memory sequences. C. R. Acad Sci Paris 333:373–376

Hurd H (1991) Correlation theory of almost periodically correlated processes. J Multivariate Anal 30:24–45

Jach A, McElroy T, Politis DN (2012) Subsampling inference for the mean of heavy-tailed long memory time series. J Time Ser Anal 33(1):96–111

Lenart L, Leśkow J, Synowiecki R (2008) Subsampling in estimation of autocovariance for PC time series. J Time Ser Anal 29(6):995–1018

McElroy T, Politis DN (2007) Self-normalization for heavy-tailed time series with long memory. Statist Sinica 17(1):199–220

Politis DN, Romano JP, Wolf M (1999) Subsampling. Springer, New York
Samorodnitsky G, Taqqu M (1994) Stable non-gaussian random processes. Chapman and Hall, New York
Taqqu M (1975) Weak convergence to fractional Brownian and to the Rosenblatt process, Zeitschrift für Wahrscheinlichkeitstheoie und verwandte Gebiete 31:287–302

Structure of PC Sequences and the 3rd Prediction Problem

Andrzej Makagon and Abolghassem Miamee

Abstract Founders of prediction theory formulated three prediction problems: extrapolation problem, interpolation problem, and the problem of positivity of the angle between the past and the future. The third one is strictly related to the question of representing the predictor as a series of past values of the process. All three have been solved in the case of stationary sequences, however, as far as we know in the case of periodically correlated sequences only the first prediction problem has been studied. The purpose of this chapter is to overview the third prediction problems.

1 Introduction

A second order stochastic sequence $(x(n))$ is a sequence of zero-mean finite variance complex random variables indexed by the set of integers $\mathscr{Z}$. At the expense of ignoring the probabilistic properties of $(x(n))$, we adopt a slightly more general definition: *a (stochastic) sequence* $(x(n))$ is a sequence in some complex, separable Hilbert space $\mathscr{H}$. By a complex, separable Hilbert space $\mathscr{H}$, we understand a vector space under the field of complex numbers $\mathscr{C}$ equipped with an inner product $(\cdot, \cdot)$ and such that admits a countable orthonormal basis.

Given a sequence $(x(n))$ in $\mathscr{H}$, the following subspaces of $\mathscr{H}$ will be of interest:

$$M_x = \overline{\mathrm{sp}}\{x(m) : m \in \mathscr{Z}\}, \ M_x(n) = \overline{\mathrm{sp}}\{x(m) : m \leq n\}, \ J_x(n) = \overline{\mathrm{sp}}\{x(m) : m \neq n\},$$

This research was supported by ARO Contract No.W911NF-11-1-0051.

A. Makagon (✉) · A. Miamee
Department of Mathematics, Hampton University, Hampton, VA 23668, USA
e-mail: andrzej.makagon@hamptonu.edu

A. Miamee
e-mail: abolghassem.miamee@hamptonu.edu

F. Chaari et al. (eds.), *Cyclostationarity: Theory and Methods*,
Lecture Notes in Mechanical Engineering, DOI: 10.1007/978-3-319-04187-2_5,
© Springer International Publishing Switzerland 2014

where $\overline{\mathrm{sp}}\{N\}$ stands for the closed linear subspace of $\mathcal{H}$ spanned by linear combinations of vectors from N. A sequence $(x(n))$ is called *regular*, if $\bigcap_n M_x(n) = \{0\}$, and is called *J-regular* if $\bigcap_n J_x(n) = \{0\}$. If M is a closed subspace of $\mathcal{H}$ and $x \in \mathcal{H}$, then $(x|M)$ will denote the orthogonal projection of x onto M; more precisely $(x|M)$ is the unique vector in M such that $x = (x|M) + e$ and $e \perp M$.

The *1st Prediction Problem* deals with regularity of a sequence $(x(n))$ and describing the projections (*predictors*) $\hat{x}(n) = (x(n)|M_x(n - 1))$ and their errors (*prediction errors*) $\sigma_n = \|x(n) - \hat{x}(n)\|$. This can be accomplished by finding the innovation of $(x(n))$. Given a sequence $(x(n))$, *the innovation of $(x(n))$* is the sequence $z_n = x(n) - \hat{x}(n)$, $n \in \mathcal{Z}$. It is clear that if $(x(n))$ is regular then $\hat{x}(n) = (x(n)|M_x(n - 1)) = \sum_{j=-\infty}^{n-1} a_j(n)z_j$ and the prediction error is $\sigma_n = \|z_n\|$. Note that the predictor is given in terms of innovations. The natural question is therefore: *Is it possible express $\hat{x}(n)$ as a convergent series $\hat{x}(n) = \sum_{k=-\infty}^{n-1} \alpha_k x(k)$ of past values of $(x(n))$?* Unfortunately, the answer to this question is *not always*, and below is a simple example in this matter.

Example 1 Let (ξ_n) be an orthonormal sequence in $\mathcal{H}$, and let $x(n) = \xi_n - \xi_{n-1}$, $n \in \mathcal{Z}$. The sequence $(x(n))$ is stationary (see Sect. 3) and regular. First we will show that (ξ_n) is the innovation for $(x(n))$. The inclusion $M_x(n) \subseteq M_\xi(n)$, $n \in \mathcal{Z}$, is obvious. To show the opposite inclusion, fix n and note that for every $m < n$, $\xi_n - \xi_m = x(n) + x(n - 1) + \cdots + x(m + 1) \in M_x(n)$. Therefore for each positive integer N

$$u_N = \xi_n - \frac{1}{N}\sum_{k=1}^{N} \xi_{n-k} = \frac{1}{N}\sum_{k=1}^{N}(\xi_n - \xi_{n-k}) \in M_x(n).$$

Consequently $\lim_N u_N = \xi_n \in M_x(n)$, which shows that $M_\xi(n) \subseteq M_x(n)$, $n \in \mathcal{Z}$. Since (ξ_n) is the innovation of $(x(n))$, the projection $\hat{x}(n) = (x(n)|M_x(n - 1))$ is equal $\hat{x}(n) = -\xi_{n-1}$. Suppose now that there is a sequence (α_k) such that

$$\hat{x}(n) = -\xi_{n-1} = \sum_{k=-\infty}^{n-1} \alpha_k x(k) = \sum_{k=-\infty}^{n-1} \alpha_k(\xi_k - \xi_{k-1}).$$

Multiplying the above by ξ_j, $j = n - 2, n - 3, \ldots$, we obtain that

$$-(\xi_{n-1}, \xi_j) = \sum_{k=-\infty}^{n-1} \alpha_k((\xi_k, \xi_j) - (\xi_{k-1}, \xi_j)) = (\alpha_j - \alpha_{j+1})\|\xi_j\|^2 = \alpha_j - \alpha_{j+1} = 0,$$

i.e. $\alpha_{n-1} = \alpha_{n-2} = \alpha_{n-3} = \cdots = \alpha$. Convergence of the series $\sum_{k=-\infty}^{n-1} \alpha_k x(k)$ implies that $\|\alpha_k x(k)\| = \sqrt{2}|\alpha_k| \to 0$ as $k \to -\infty$, and hence $\alpha = \alpha_{n-1} = \alpha_{n-2} = \alpha_{n-3} = \cdots = 0$, which leads to a contradiction.

We elaborate for a while on this example and look at the *2nd prediction problem*: the problem of J-regularity and predicting a value $x(n)$ given all other values $x(m)$, $m \neq n$, of the sequence. The sequence $(x(n))$ in Example 1 admits an exact prediction because for this sequence $J_x(n) = M_x$ for all $n \in \mathcal{Z}$. To see this suppose that $y \in M_x$

is orthogonal to each $x(m)$, $m \neq n$, i.e. $(y, \xi_m - \xi_{m-1}) = 0$ for every $m \neq n$. Since (ξ_k) is an orthonormal basis for M_x, $y = \sum_{j=-\infty}^{\infty} c_j(y)\xi_j$. Multiplying the last equality by $\xi_m - \xi_{m-1}$, $m \neq n$, we obtain that

$$\sum_{j=-\infty}^{\infty} c_j(y)(\xi_j, \xi_m - \xi_{m-1}) = c_m(y) - c_{m+1}(y) = 0 \quad \text{for every } m \neq n.$$

Therefore all $c_k(y)$, $k \in \mathcal{Z}$, are equal, and because the series $\sum_{j=-\infty}^{\infty} c_j(y)\xi_j$ converges, all must be equal 0. Hence $y = 0$.

In fact, not the lack of J-regularity but the fact that the angle between the past and the future for this sequence is zero, is the real reason of absence of autoregressive representation for the predictors. This bring us the the *3rd prediction problem*: the question of positivity of the angle between the past and the future. Given two nonzero closed subspaces P and F of a Hilbert space $\mathcal{H}$, the *cosine of the angle* between P and F is defined as

$$\cos \angle(P, F) = \sup\{|(x, y)| : x \in P, y \in F, ||x|| = ||y|| = 1\} \tag{1}$$

If $\cos \angle(P, F) = 1$ then the angle $\angle(P, F)$ between P and F is said to be 0, otherwise we say that the angle between P and F is *positive*. It is easy to see that $\angle(P, F) = 0$ iff there is a positive constant $\delta > 0$ and two sequences $p_N \in P$ and $f_N \in F$ such that for sufficiently large N, $||p_N|| \geq \delta$, $||f_N|| \geq \delta$, and $||p_N - f_N|| \to 0$. This follows readily from the *Law of Cosines*

$$||x - y||^2 = ||x||^2 + ||y||^2 - 2\Re(x, y) \geq ||x||^2 + ||x||^2 - 2|(x, y)|. \tag{2}$$

Let $P_0 = \overline{\text{sp}}\{x(k) : k \leq 0\}$ and $F_0 = \overline{\text{sp}}\{x(k) : k > 0\}$ be the past and present and the future of $(x(n))$ with regard to the a time moment $t = 0$. We will show that for the sequence $x(n) = \xi_n - \xi_{n-1}$ from Example 1, the angle between P_0 and F_0 is 0. Since $x(0) \in J_x(0)$, there is a sequence S_N of finite linear combinations of vectors $x(m)$, $m \neq 0$, that converges to $x(0)$ as $N \to \infty$

$$S_N = \sum_{m \neq 0} c_{N,m} x(m) = \sum_{m < 0} c_{N,m} x(m) + \sum_{m > 0} c_{N,m} x(m) \longrightarrow x(0).$$

Denoting the sums above by S_N^- and S_N^+, respectively, we obtain that $S_N^+ - (x(0) - S_N^-) \to 0$, which in particular implies that $||S_N^+|| - ||(x(0) - S_N^-)|| \to 0$. Since $(x(n))$ is regular, $||x(0) - S_N^-|| \geq ||x(0) - \hat{x}(0)|| = 1 > 1/2$ for all N, and hence also $||S_N^+|| > 1/2$ for sufficiently large N. Therefore the sequences $p_N = (x(0) - S_N^-)$ and $f_N = S_N^+$ satisfy the conditions of the above criterion, and hence the angle between P_0 and F_0 is 0. The reader has probably noticed that we made the proof more complicated than it is. Repeating the argument used at the beginning of the

Example 1, we see that ξ_0 belongs to both P_0 and F_0, so clearly $\angle(P_0, F_0) = 0$. However, there exist examples for which $\angle(P_0, F_0) = 0$ although $P_0 \cap F_0 = \{0\}$. $\square$

The chapter has three parts. In Sect. 2 we review some facts about angles and bases. Section 3 contains a sketch of the theory of periodically correlated (PC) sequences. Our presentation is different than the standard (as presented for example in Hurd and Miamee 2007) and is based on a description of the structure of PC sequences which is given in Theorem 3.2. In Sect. 4 we give a certain sufficient condition for a periodically correlated sequence to be of positive angle.

2 Angles and Bases

We will have to talk about linear operators on a Hilbert space $\mathscr{H}$. A mapping $A : \mathscr{H} \to \mathscr{H}$ is called an *operator* (or *linear operator*) if it satisfies $A(ax + by) = aA(x) + bA(y)$, for all scalars a, b and vectors $x, y \in \mathscr{H}$. It is customary to skip parentheses around (x), so we often write Ax instead of $A(x)$. An operator A is continuous in the topology induced by a norm on $\mathscr{H}$ if and only if A is *bounded*, that is

$$\|A\| = \sup\{\|Ax\|; \|x\| \le 1\} < \infty$$

Hence A is continuous $(=$ bounded$)$ iff there is a constant $C < \infty$ such that $\|Ax\| \le C\|x\|$, for all $x \in \mathscr{H}$. The smallest C above is of course equal to $\|A\|$. The *Closed Graph Theorem* states that A is bounded iff its graph graph$(A) = \{(w, Aw) : w \in \mathscr{H}\}$ is a closed subset of $\mathscr{H} \times \mathscr{H}$, that is iff for every sequence w_n in $\mathscr{H}$ the following implication is true: if $w_n \to w$ and $Aw_n \to u$, then $u = Aw$. If $\mathscr{H}_0$ is a linear dense subset of $\mathscr{H}$ and A is a linear function $A : \mathscr{H}_0 \to \mathscr{H}$ such that there is a $C < \infty$ with the property that $\|Ax\| \le C\|x\|$ for every $x \in \mathscr{H}_0$, then A has a unique extension to a bounded operator on $\mathscr{H}$. This is a consequence of completeness of $\mathscr{H}$. A linear function $c : \mathscr{H} \to \mathscr{C}$ from $\mathscr{H}$ to the set of complex number $\mathscr{C}$ is called a *linear functional*. A linear functional c is continuous iff it is bounded, that is iff there is a constant $C < \infty$ such that $|c(x)| \le C\|x\|$ for all $x \in \mathscr{H}$. The *Riesz Representation Theorem* states that c is a bounded linear functional on a Hilbert space $\mathscr{H}$ iff there is a unique $h \in \mathscr{H}$ such that $c(x) = (x, h), x \in \mathscr{H}$.

We start with angles.

Proposition 2.1 *Let M and N be two nonzero subspaces of $\mathscr{H}$ such that $\overline{sp}\{M + N\} = \mathscr{H}$. The following conditions are equivalent:*

1. *$\angle(M, N) > 0$,*
2. *there exists a constant $C < \infty$ such that for every $x \in M$ and $y \in N$, $\|x\| \le C\|x + y\|$,*
3. *every $w \in \mathscr{H}$ can be in the unique way written as $w = x + y$ where $x \in M$ and $y \in N$ (this property says that $M + N = \mathscr{H}$.)*

Moreover, the smallest constant C satisfying inequality in condition 2. equals $1/(2 - 2\rho)$ where $\rho = \cos\angle(M, N)$.

Proof The proof is a textbook linear algebra.

(1. $\Rightarrow$ 2.) First suppose that $\angle(M, N) = \rho > 0$. Let $x \in M$ and $y \in N$. The inequality is true if $x = 0$. Assume for a moment that $\|x\| = 1$ and write $y = az$, where $\|z\| = 1$. Then from (2) it follows that $\|x + az\|^2 \geq 1 + |a|^2 - 2|a|\rho = f(|a|)$. The minimum of the quadratic function $f(t) = 1 + t^2 - 2t\rho$, $t \in \mathcal{R}$, is attained at $t = \rho$ and is equal $1 - \rho^2$. Hence $\|x + y\|^2 \geq 1 - \rho^2$. If $\|x\| \neq 1$, then

$$\|x + y\| = \|x\|\|(x/\|x\|) + (y/\|x\|)\|\|x\| \geq \|x\|\sqrt{1 - \rho^2},$$

Therefore we obtain that $\|x\| \leq C\|x + y\|$ with $C = \sqrt{1/(1 - \rho^2)}$. Note that $\|y\| = \|y + x - x\| \leq \|y + x\| + \|x\| \leq (C + 1)\|x + y\|$, so also $\|y\| \leq C\|x + y\|$, for some (possibly different) $C < \infty$.

(2. $\Rightarrow$ 3.) The inequality 2. implies that $M \cap N = \{0\}$. Indeed, if $v \in M \cap N$, then $\|v\| = C\|v + (-v)\| = 0$, so $v = 0$. Let $w \in \mathcal{H}$. Since $\overline{\mathrm{sp}}\{M + N\} = \mathcal{H}$, there there exist sequences $x_n \in N$ and $y_n \in M$, such that $w = \lim_n(x_n + y_n)$. Therefore $w_n = x_n + y_n$ is Cauchy. By 2. both (x_n) and (y_n) are Cauchy and hence converge to, say, x and y, respectively. Since both N and M are closed, $x \in N$ and $y \in M$. Therefore $w = x + y$. The representation is unique since, if $w = x + y = u + v$, $x, y \in N, y, v \in M$ then $x - y = v - y$ belongs to both N and M, so $x - y = v - y = 0$.

(3. $\Rightarrow$ 1.) Suppose that for every every $w \in \mathcal{H}$ there are unique $x \in M$ and $y \in N$ such then $w = x + y$. Then the mapping $P_M(w) = x$ is well defined linear operator from $\mathcal{H}$ into $\mathcal{H}$. The operator P_M has a closed graph. To show this suppose that $w_n = x_n + y_n \to w = x + y$, $x \in M$, $y \in N$, and that $P_M(w_n) = x_n \to u \in M$. Consequently $y_n = w_n - x_n \to v = w - u \in N$. Hence $w = x + y = u + v$, and from the uniqueness of the decomposition of w, we conclude that $u = x = P_M(w)$. From the Closed Graph Theorem it therefore follows that P_M is bounded, that is $\|P_M(x + y)\| = \|x\| \leq C\|x + y\|$, for all $x \in M$ and $y \in N$, where $C = \|P_M\|$. To show that $\angle(M, N) > 0$, suppose on the contrary that $\angle(M, N) = 0$. From the Low of Cosines (2) it follows that there exist sequences $x_n \in M$ and $y_m \in M$ with $\|x_n\| = \|y_n\| = 1$ such that $\|x_n - y_n\| \to 0$. The boundedness of P_M, however, implies that $\|x_n\| \leq C\|x_n - y_n\| \to 0$. This contradicts the fact that for every n, $\|x_n\| = 1$. We skip the proof on the *moreover* part. $\square$

The operator $P_M(x + y) = x$, $x \in M$, $y \in N$, defined in the proof above is called the *projection onto M along N*. We just have proved that P_M is bounded iff $\angle(M, N) > 0$.

Definition 2.1 *Let $(x(n))$ be a sequence in $\mathcal{H}$ and let for every $n \in \mathcal{Z}$ the past and present P_n and the future F_n of the sequence $(x(n))$ at a time moment n be defiend as*

$$P_n = M_x(n) = \overline{\mathrm{sp}}\{x(t) : t \leq n\} \quad and \quad F_n = \overline{\mathrm{sp}}\{x(t) : t > n\}. \tag{3}$$

We say that $(x(n))$ is of positive angle *if* $\sup(\cos\angle(P_n, F_n)) < 1$.

From Proposition 2.1 it follows that $(x(n))$ is of positive angle iff for every n there is C_n such that $\|x\| \le C_n \|x + y\|$ for all $x \in P_n$ and $y \in F_n$ (Lemma 2.1). If $(x(n))$ is of positive angle and all $x(n) \ne 0$, then $(x(n))$ must be linearly independent; otherwise we would find $n_0 < n_1 < \cdots < n_p$ and nonzero a_k, $k = 1, \ldots, p$, such that $x(n_0) = \sum_{k=1}^{p} a_k x(n_k)$ and the vector $x(n_0)$ then would belong to both P_{n_0} and F_{n_0}. Also note that if $(x(n))$ is of positive angle then for every two sets Δ_1 and Δ_2 such that $\Delta_1 \subseteq (-\infty, n]$ and $\Delta_2 \subseteq (n, \infty)$ for some $n \in \mathscr{Z}$, the angle between $M_x(\Delta_1) = \overline{\mathrm{sp}}\{x(k) : k \in \Delta_1\}$ and $M_x(\Delta_2)$ is positive, provided that both are nonzero. Less obvious is the fact that if $(x(n))$ is of positive angle then for every $m \le n$ the angle between $M_x((m, n))$ and $M_x((m, n)^c)$ is positive, provided $M_x((m, n)) \ne \{0\}$. To see this let us assume that $x \in M_x((m, n))$ and $y \in M_x((m, n)^c)$. Because $\angle(M_x((-\infty, m)), M_x((n, \infty)) > 0$, from Proposition 2.1 it follows that $y = u + v$, where $u \in M_x((-\infty, m))$ and $v \in M_x((n, \infty))$. Since $x \in P_n$ and $v \in F_n$ we have $\|x\| \le C_n \|x + v\|$; furthermore, since $x + v \in F_{m-1}$ and $u \in P_{m-1}$, we have $\|x + v\| \le C_{m-1}\|x + v + u\|$. Hence $\|x\| \le C_n C_{m-1}\|x + v + u\| = C_{m,n}\|x + y\|$. This implies that if $(x(n))$ is of positive angle and all $x(n) \ne 0$, then all projection $P_{m,n}$ onto $M_x((m, n))$ along $M_x((m, n)^c)$, $m \le n$, are uniformly bounded.

The last property of sequences with positive angle links us with the notion of a basis.

Definition 2.2 *A sequence* $(x(n))$, $n \in \mathscr{Z}$, *in a Hilbert space* $\mathscr{H}$ *is called a (Schauder) basis for* $\mathscr{H}$ *if for every* $x \in \mathscr{H}$ *there is a unique sequence of scalars* $(c_n(x))$, *such that*

$$x = \sum_{n=-\infty}^{\infty} c_n(x)x(n), \tag{4}$$

A basis $(x(n))$ *is called* unconditional *if the series (4) converges unconditionally (i.e. for every permutation of* $\mathscr{Z}$ *or, equivalently, for any sequence of signs* $\pm$ *in front of coefficients* c_n*).*

Here $\sum_{n=-\infty}^{\infty} c_n(x)x(n) = \lim_{M \to -\infty, N \to \infty} \sum_{n=M}^{N} c_n(x)x(n)$. Every orthonormal basis for $\mathscr{H}$ is an unconditional Schauder basis for $\mathscr{H}$. Note that the fact that the sequence $(c_n(x))$ is unique implies that $x(n) \ne 0$ for every $n \in \mathscr{Z}$, and that the coefficients $c_n(x)$ are linear functions of x.

The proposition below belongs to Banach and is the starting point to the vast theory of Schauder bases.

Proposition 2.2 (E.g. Lindenstrauss and Tzafiri (1997), Prop 1.a.3) *Let* $(x(n))$ *be a sequence in* $\mathscr{H}$*. Then* $(x(n))$ *is a basis for* $\mathscr{H}$ *iff the following three conditions hold:*

1. $x(n) \ne 0$ *for all* n,
2. $\overline{\mathrm{sp}}\{x(k) : k \in \mathscr{Z}\} = \mathscr{H}$,

3. *there is a constant $K > 0$ such that for any choice of scalars (a_n) and for any integers $-\infty < p \le m \le n \le q < \infty$*

$$\left\| \sum_{k=m}^{n} a_k x(k) \right\| \le K \left\| \sum_{k=p}^{q} a_k x(k) \right\|, \tag{5}$$

For the proof please see Lindenstrauss and Tzafiri (1997) or Heil (2011). If $(x(n))$ is a basis for $\mathcal{H}$ then by taking $n = m$ in (5) we obtain that each $c_n(x)$ in (4) is a continuous linear functional on $\mathcal{H}$. The Riesz Representation Theorem implies that there exist a sequence $(h(n))$ in $\mathcal{H}$ such that $c_n(x) = (x, h(n))$, $n \in \mathcal{Z}$. The sequence $(h(n))$ is unique because $c_n(x)$ are so. From the uniqueness of the representation (4) it also follows that $c_n(x(k)) = (x(k), h(n)) = \delta_{k,n}$, that is for every $n \in \mathcal{Z}$, $h(n) \perp J_x(n)$. Sequences $(x(n))$ and $(h(n))$ such that $(x(k), h(n)) = \delta_{k,n}$ are called *biorthogonal*. Here is a summary of our findings.

Corollary 2.1 *Suppose that $x(n) \ne 0$ for all n. The following three statement about the sequence $(x(n))$ are equivalent.*

1. *$(x(n))$ is of positive angle.*
2. *$(x(n))$ is a basis for M_x.*
3. *There is a sequence $(h(n))$ in M_x biorthogonal to $(x(n))$ such that for every $x \in M_x$.*

$$x = \sum_{n=-\infty}^{\infty} (x, h(n))x(n). \tag{6}$$

Moreover, if $(x(n))$ is a basis for M_x, then $(h(n))$ is also a basis for M_x.

Proof $(1. \Leftrightarrow 2.)$ If $x(n) \ne 0$, $n \in \mathcal{Z}$, then $M_x((m, n))$ is nonzero for every $m \le n$ and, as we have already noticed, all projections projections $P_{m,n}$ are well defined and uniformly bounded, which is exactly the necessary and sufficient condition 3. for $(x(n))$ to be a basis in M_x. Conversely, let $n \in \mathcal{Z}$ and let $v = x + y$, where $x \in P_n$ and $y \in F_n$. Let x_m and y_m be a sequences of finite linear combinations of $x(k)$, $k \le n$ and $x(k)$, $k > n$, respectively, that converge to x and y. From (5) it follows that that there is a constant $K < \infty$, which does not depend on n, such that for $\|x_m\| \le K \|x_m + y_m\|$ for all m. Therefore $\|x\| \le K \|x + y\|$. Relation established in Proposition 2.1 implies that for every $n \in \mathcal{Z}$, $\cos \angle(P_n, F_n) < 1 - 1/2K < 1$.

$(2. \Leftrightarrow 3.)$ The existence of $(h(n))$ has already been mentioned. For converse we only need to show that the representation (6) is unique. Suppose that $x = \sum_{n=-\infty}^{\infty} c_n x(n)$. Then, because of biorthogonality, $(x, h(k)) = \sum_{n=-\infty}^{\infty} c_n(x(n), h(k)) = c_k$. For the proof of the *moreover* part we refer to Heil (2011). $\square$

None of the results above sections are new. Facts about Schauder basis are from Lindenstrauss and Tzafiri (1997); Heil (2011), facts about angles can be found in Helson and Szegö (1960) or Pourahmadi (2001). For another approach to the problem of basis please see Miamee (1991, 1993).

In a process of studying prediction problems, a series of related properties of a sequence $(x(n))$ have been singled out. Here are some of them. A sequence $(x(n))$ in a Hilbert space $\mathcal{H}$ is said to be:

> *regular* (r), if $\bigcap_n M_x(n) = \{0\}$;
> *full rank regular* (frr) if its regular and for every $n \in \mathcal{Z}$, $x(n) \notin M_x(n-1)$;
> *J-regular* (Jr) if $\bigcap_n J_x(n) = \{0\}$;
> *J-full rank regular* (Jfrr) if its J-regular and for every $n \in \mathcal{Z}$, $x(n) \notin J_x(n)$;
> *a basis* (b) if $(x(n))$ is a Schauder basis for M_x;
> *of positive angle* (pa) if it satisfies the conditions of Definition 2.1.

Moreover $(x(n))$ is said to have a *moving average representation* (ma) if there is an orthonormal sequence (ξ_n) in $\mathcal{H}$ such that for every $n \in \mathcal{Z}$ there exists a sequence $\theta_k(n), k = 0, 1, \ldots,$ such that

$$x(n) = \sum_{k=0}^{\infty} \theta_k(n)\xi_{n-k}; \tag{7}$$

and is said to have an *autoregressive representation* (ar) if $(x(n))$ is regular and for every $n \in \mathcal{Z}$ there exists a unique sequence $\phi_k(n), k = 1, 2, \ldots$ such that

$$x(n) = z_n + \sum_{k=1}^{\infty} \phi_k(n)x(n-k),$$

where (z_n) is the innovation of $(x(n))$.

Being of positive angle is the strongest of the above properties.

Proposition 2.3 *Suppose that for all $n \in \mathcal{Z}$, $x(n) \neq 0$. Then*

$$(\text{pa}) \Leftrightarrow (\text{b}) \Rightarrow (\text{Jfrr}) \Rightarrow (\text{frr}) \Rightarrow (\text{r}) \Leftrightarrow (\text{ma}) \tag{8}$$

$$(\text{pa}) \Rightarrow (\text{ar}) \tag{9}$$

None of the one-sided implications in (8) can be reversed.

Proof Most of the implications are either obvious or have already been proved. (pa) $\Leftrightarrow$ (b) is included in Corollary 2.1. To prove the implication (b) $\Rightarrow$ (Jfrr) note that by Corollary 2.1, the vectors $h(n)$ are in $M_x \ominus J_x(n)$ and they form a basis for M_x. To see that (Jfrr) $\Rightarrow$ (frr) note that $M_x(n) \subseteq J_x(n)$, so (Jr) implies (r). If additionally $x(n) \notin J_x(n)$, then $x(n) \notin M_x(n) \subseteq J_x(n)$. The other implications are obvious. Fallacy of implication (r) $\Rightarrow$ (Jr) is exhibited in Example 1. The fallacy of (Jmr) $\Rightarrow$ (b) was pointed by Babenko (1948). We do not know whether (ar) implies (pa). $\qquad\qquad\square$

3 Structure of Periodically Correlated Sequences and Its Consequences

If $(x(n))$ is a sequence in a Hilbert space $\mathscr{H}$, then *the covariance function* R_x of $(x(n))$ is the function on $\mathscr{Z} \times \mathscr{Z}$ defined by $R_x(n, m) = (x(n), x(m))$, where $(\cdot, \cdot)$ is the inner product in $\mathscr{H}$. Two sequences $(x(n))$ and $(y(n))$, in possibly different Hilbert spaces $\mathscr{H}$ and $\mathscr{K}$, are said to be *unitary equivalent* if $R_x(m, n) = R_y(m, n)$ for every $m, n \in \mathscr{Z}$. Unitary equivalent sequences will be identified. This identification makes the space $\mathscr{H}$ where the values of $(x(n))$ are physically located completely irrelevant.

Before we proceed we need to say a few words about groups, measures, and unitary operators. The set of integers $\mathscr{Z}$ is an additive group. The set $\hat{\mathscr{Z}}$ of all homomorphisms from $\mathscr{Z}$ into the multiplicative group $\{z \in \mathscr{C} : |z| = 1\}$ is called the dual of $\mathscr{Z}$. Under the operation of composition, the dual $\hat{\mathscr{Z}}$ is a group itself. The group $\hat{\mathscr{Z}}$ can be identified with the interval $[0, 2\pi)$ with addition modulo 2π; in the future when we write $[0, 2\pi)$ we would always think about it as the dual of $\mathscr{Z}$. A measure μ on $[0, 2\pi)$ is a complex σ-additive set function defined on the Borel σ-algebra $\mathscr{B}$ of $[0, 2\pi)$. A measure μ is nonnegative if $\mu(\Delta) \geq 0$ for all $\Delta \in \mathscr{B}$. If μ is a nonnegative measure on $[0, 2\pi)$ then $L^2(\mu; \mathscr{C}^d)$ will denote the set of all $\mathscr{C}^d$-valued μ-measurable functions on $[0, 2\pi)$ such that $\int_0^{2\pi} \|f(t)\|^2 \mu(dt) < \infty$. Equipped with the inner product

$$(f, g) = \int_0^{2\pi} f(t)g(t)^* \mu(dt),$$

$L^2(\mu; \mathscr{C}^d)$ becomes a Hilbert space. Here and in the sequel an element $a \in \mathscr{C}^d$ will be represented as a row-vector and a^* will stand for a column-vector whose coordinates are complex conjugates of the coordinates of a. A measure μ on $[0, 2\pi)$ is called *absolutely continuous* if there is an integrable function f such that for every $\Delta \in \mathscr{B}$, $\mu(\Delta) = \int_\Delta f(t)dt$, where dt denotes the Lebesgue measure. The function $f(t)$ (if exists) is called the density of μ and will be denoted $\frac{d\mu}{dt}(t)$ or $\mu'(t)$. A mesure μ on $[0, 2\pi)$ is called *a-invariant*, $a \in [0, 2\pi)$, if $\mu(\Delta + a) = \mu(\Delta)$ for every $\Delta \in \mathscr{B}$.

Given two Hilbert spaces $\mathscr{H}$ and $\mathscr{K}$, a linear continuous transformation $U : \mathscr{H} \to \mathscr{K}$ is called an *unitary operator* if U is onto and for every $x, y \in \mathscr{H}$, $(Ux, Uy) = (x, y)$. Note that two sequences $(x(n))$ and $(y(n))$ are unitary equivalent iff there is a unitary $\Phi : M_x \to M_y$ such that $\Phi(x(n)) = y(n), n \in \mathscr{Z}$. The Spectral Theorem says that if U is a unitary operator from a Hilbert space $\mathscr{H}$ onto itself, then there exists a unique set function E defined on $\mathscr{B}$ such that for every $x, y \in \mathscr{H}$, $(E(\cdot)x, y)$ is a measure, $E(\Delta)$ is an orthogonal projection in $\mathscr{H}$, $E(\Delta_1)E(\Delta_2) = E(\Delta_1 \cap \Delta_2)$, and such that for every $n \in \mathscr{Z}$

$$U^n x = \int_0^{2\pi} e^{-iun} E(du)x, \quad x \in \mathcal{H}. \tag{10}$$

The measure E is called a *resolution of identity of U*.

3.1 Stationary Sequences

The simplest periodically correlated sequences are stationary. A sequence $(x(n))$ is said to be *stationary*, if $R_x(m+1, n+1) = R_x(m, n)$ for every $m, n \in \mathscr{Z}$. Stationary sequences have a very simple structure.

Theorem 3.1 *(Structure od a Stationary Sequence) $(x(n))$ is stationary iff there is a Hilbert space $\mathscr{K} \supseteq M_x$, a unitary operator U in $\mathscr{K}$, and $x \in \mathscr{K}$, such that*

$$x(n) = U^n x, \quad n \in \mathscr{Z}. \tag{11}$$

This theorem is a starting point to a diverse and beautiful prediction theory of stationary sequences as developed Kolmogorov, Wiener, Helson, Szego, Masani, Rozanow, Urbanik and others. First note that from (10) it follows that every stationary sequence has an integral representation $x(n) = \int_0^{2\pi} e^{-inu} E(du)x, n \in \mathscr{Z}$. Consequently

$$R_x(m, n) = \int_0^{2\pi} e^{-i(m-n)u} \gamma(du), \quad \text{where } \gamma(du) = (E(du)x, x). \tag{12}$$

The nonnegative measure γ above is uniquely determined by R_x and is called *the spectral measure* of a stationary sequence $(x(n))$. If γ is absolutely continuous with respect a nonnegative measure μ and $h \in L^2(\mu, \mathscr{C})$ is such that $\frac{d\gamma}{d\mu}(t) = |h(t)|^2$, dt−a.e., then from (12) it follows that the sequence $(f(n))$ of functions in $L^2(\mu, \mathscr{C})$ defined as $f(n)(u) = e^{-inu} h(u), n \in \mathscr{Z}$, has the same covariance function as $(x(n))$ and hence the sequences $(x(n))$ and $(f(n))$ are unitary equivalent. This functional representation of $(x(n))$ connects the prediction problem with Harmonic Analysis, and in particular with the theory of invariant subspaces, Helson (1964).

All three prediction problems are solved for stationary sequences. A concise summary of prediction theory of stationary sequences, including solutions to prediction problems, can be found in last chapters of Pourahmadi (2001). The prediction theory was extended to finite dimensional stationary sequences (Makagon and Weron 1976; Masani 1960, 1966; Wiener and Masani 1957, 1958), and then to infinite dimensional stationary sequences (see Makagon and Salehi 1989 for summary and references). However, in the multidimensional case only the 2nd prediction problem has satisfactory solution (Makagon 1984), the other two have only partial solutions.

A *T-dimensional stationary sequence* is a family $(x^k(n))$, $k = 1, \ldots, T$, of T stationary sequences which are stationary cross-correlated, that is such that for every $n, m \in \mathscr{Z}$ and $k, j = 1, \ldots, T$

$$R^{j,k}(m,n) := (x^j(m), x^k(n)) = (x^j(m+1), x^k(n+1)) = R^{j,k}(m+1, n+1).$$

The past of a T-dimensional stationary sequence is defined as $M_x(n) = \overline{\mathrm{sp}}\{x^k(m) : m \le n, k = 1, \ldots, T\}$. All components of a T-dimensional stationary sequence have the same shift operator U, that is $x^k(n) = U^n x^k(0)$ for all $n \in \mathscr{Z}$ and $k, j = 1, \ldots, T$. Consequently for each $k, j = 1, \ldots, T$ there exists a measure $\gamma^{j,k}$ m on $[0, 2\pi)$ such that

$$R^{j,k}(m,n) = \int_0^{2\pi} e^{-i(m-n)t} \gamma^{j,k}(dt), \quad m, n \in Z.$$

The $T \times T$ - matrix measure $\Gamma(\Delta)$ whose (j, k) entry is $\gamma^{j,k}(\Delta)$ is called the spectral measure if a T-dimensional stationary sequence $(x(n))$.

3.2 Periodically Correlated Sequences

Let T be a positive integer. A sequence $(x(n))$ is called *periodically correlated* (PC) with period T, or simply T-PC, if $R_x(m+T, n+T) = R_x(m, n)$ for every $m, n \in \mathscr{Z}$.

Following the development of the prediction theory for stationary sequences, we start with a description of the structure of T-PC sequences.

Theorem 3.2 (Structure of a T-PC Sequence) *A sequence $(x(n))$ is T-PC iff there is a Hilbert space $\mathscr{K} \supseteq M_x$, unitary operators U and V in $\mathscr{K}$ such that $V^T = I$ and $VU = e^{-2\pi i/T} UV$, and an $x \in \mathscr{K}$ such that*

$$x(n) = (1/T) \sum_{j=0}^{T-1} e^{-2\pi i jn/T} U^n V^j x, \quad n \in \mathscr{Z}. \tag{13}$$

A noticeable difference between structure of a stationary sequences and the structure of T-PC sequences (with $T > 1$) is that in Theorem 3.1 the space $\mathscr{K}$ could be chosen to be M_x, while in Theorem 3.2 the space $\mathscr{K}$ is larger than M_x. The proof is a straightforward application of Mackey's inducing construction (cf. Folland 1995, Chapter 6). Also it can be found between lines of Makagon (2011), although in the latter chapter the theorem is stated in a different way.

Proof It is clear that if $(x(n))$ satisfies (13) then $R_x(n + T, m + T) = R_x(n, m)$, so $(x(n))$ is T-PC. Conversely, suppose that $(x(n))$ is T-PC. The sequence $(x(n))$ generates the the unitary operator W in M_x (so called T-shift) via the equation $Wx(k) = x(k + T), k \in \mathscr{Z}$. Let $\mathscr{K} = M_x^{\oplus T}$. Define two operators U and V in $\mathscr{K}$ by

$$U(x_1, x_2, \ldots, x_T) = (x_2, x_3, \ldots, Wx_1),$$
$$V(x_1, x_2, \ldots, x_T) = (x_1, x_2 e^{2\pi i/T}, \ldots, x_T e^{2\pi i(T-1)/T})$$

Then $UV(x_1, x_2, \ldots, x_T) = (x_2 e^{2\pi i/T}, \ldots, x_T e^{2\pi(T-1)i/T}, Wx_1)$, while $VU(x_1, x_2, \ldots, x_T) = V(x_2, \ldots, x_T, Wx_1) = (x_2, \ldots, x_T e^{2\pi i(T-2)/T}, Wx_1 e^{2\pi i(T-1)/T})$. Hence $VU = e^{-2\pi i/T} UV$ and more generally

$$V^j U^n = e^{-2\pi i n j/T} U^n V^j. \tag{14}$$

Clearly $V^T(x_1, x_2, \ldots, x_T) = (x_1, x_2 e^{i2\pi}, \ldots, x_T e^{i2\pi(T-1)}) = (x_1, x_2, \ldots, x_T)$. Also note that $\sum_{j=0}^{T-1} V^j(x_1, x_2, \ldots, x_T) = \sum_{j=0}^{T-1}(x_1, x_2 e^{2\pi ij/T}, \ldots, x_T e^{2\pi i(T-1)j/T}) = (Tx_1, 0, \ldots, 0)$.

Let $x = (x(0), x(1), \ldots, x(T-1))$. From the above equality, the commutativity property (14), and the fact that $U^n x = (x(n), x(n+1), \ldots, x(n+T-1))$ we conclude that

$$(1/T) \sum_{j=0}^{T-1} e^{-2\pi ijn/T} U^n V^j x = (1/T) \left(\sum_{j=0}^{T-1} V^j \right) U^n x = (x(n), 0, \ldots, 0), \tag{15}$$

the latter being obviously unitary equivalent to $(x(n))$. $\qquad\square$

The Eq. (13) implies that each T-PC sequence can be written in the form $x(n) = U^n p(n)$, where $p(n) = (1/T) \sum_{j=0}^{T-1} e^{-2\pi ijn/T} V^j$ is a T-periodic sequence in $\mathcal{K}$. The fact that each T-PC sequence is a unitary deformation of a periodic function is known and was first proved by Hurd. The representation $x(n) = U^n p(n)$ is, however, not unique in general. An advantage of our representation (13) is that U, V and x are unique up to unitary equivalence.

Proposition 3.1 *Suppose that*

$$x_k(n) = \frac{1}{T} \sum_{j=0}^{T-1} e^{-2\pi ijn/T} U_k^n V_k^j x_k, \quad n \in \mathscr{Z}, k = 1, 2,$$

where U_k and V_k are unitary operators in $\mathscr{K}_k$ such that $V_k^T = I$, $V_k U_k = e^{-2\pi i/T} U_k V_k$, and $x_k \in \mathscr{K}_k$, $k = 1, 2$. Assume additionally that $\mathscr{K}_k = \overline{sp}\{V_k^j U_k^n x_k : j, n \in \mathscr{Z}\}$, $k = 1, 2$. Then the sequences $(x_1(n))$ and $(x_2(n))$ are unitary equivalent iff there exists a unitary operator $\Phi : \mathscr{K}_1 \to \mathscr{K}_2$ such that $\Phi x_1 = x_2$, $\Phi U_1 = U_2 \Phi$ and $\Phi V_1 = V_2 \Phi$.

Proof Denote $R^k(n+r, r) = (x_k(n+r), x_k(r))$, and let $a_p^k(n) = \sum_{r=0}^{T-1} e^{-2\pi ipr/T} R^k(n+r, r)$, $k = 1, 2$. For every $n \in \mathscr{Z}$, the sequence $a_p^k(n)$, $p \in \mathscr{Z}$, is a discrete Fourier transform of the T-periodic sequence $R^k(n+r, r)$, $r \in \mathscr{Z}$, and hence $R^1 = R^2$ iff $a^1 = a^2$. The commutation relation implies that

$$R^k(n+r, r) = \frac{1}{T^2} \sum_{j=0}^{T-1} \sum_{p=0}^{T-1} e^{-2\pi i r(j-p)/T} (V_k^j U_k^n x_k, V_k^p x_k),$$

and hence for every $n, q \in \mathscr{Z}$

$$a_q^k(n) = \frac{1}{T^2} \sum_{j=0}^{T-1} \sum_{p=0}^{T-1} \left(\sum_{r=0}^{T-1} e^{-2\pi i r(j-p+q)/T} \right) (U_k^n x_k, V_k^{p-j} x_k) = (U_k^n x_k, V_k^q x_k).$$

In particular, for every $m, n, p, j \in \mathscr{Z}$ and $k = 1, 2$ we have that

$$(U_k^m V_k^p x_k, U_k^n V_k^j x_k) = e^{2\pi i (m-n)p/T} a_{j-p}^k(m-n). \tag{16}$$

If $(x_1(n))$ and $(x_2(n))$ are unitary equivalent, then $R^1 = R^2, a_r^1(n) = a_r^2(n)$ for every $r, n \in \mathscr{Z}$, and hence the inner product above does not depend on k. Consequently the mapping $\Phi\left((U_1^n)V_1^j x_1\right) = U_2^n V_2^j x_2$ is well defined and extends linearly to an isometry from $\mathscr{K}_1$ onto $\mathscr{K}_2$. This proves the "only if" part. The "if" part is obvious.
$\square$

Note that due to canonical commutation relation, the Eq. (13) can be written as

$$x(n) = \left(\frac{1}{T} \sum_{j=0}^{T-1} V^j \right) U(n)x = P_0 y(n), \quad n \in \mathscr{Z},$$

where $P_0 = (1/T) \sum_{j=0}^{T-1} V^j$ is the orthogonal projection in $\mathscr{K} = M_x^{\oplus T}$ onto the first coordinate and $y(n) = U(n)x$ is a stationary sequence in $\mathscr{K}$. Hence, as a by-product, we have constructed an explicit *dilation* of a T-PC sequence to a stationary sequence.

If U, V are two unitary operators in a Hilbert space $\mathscr{K}$ and $V^T = I$, then the powers $U^n, n \in \mathscr{Z}$, and $V^j, j = 0, \ldots, T-1$, form unitary representations of $\mathscr{Z}$, and the quotient group $\mathscr{Z}_T = \mathscr{Z}/T\mathscr{Z} = \{0, \ldots, T-1\}$, respectively. The spectral theorem for unitary group representations implies that there exist a resolution of identity E on $[0, 2\pi)$ and a resolution of identity P on $\widehat{\mathscr{Z}_T} = \{0, 2\pi/T, 4\pi/T, \ldots, 2(T-1)\pi/T\}$ (which is regarded as a subgroup of the circle $[0, 2\pi)$, i.e. the addition in $\widehat{\mathscr{Z}_T}$ is modulo 2π) such that

$$U^n = \int_0^{2\pi} e^{-iun} E(du), \quad n \in \mathscr{Z} \tag{17}$$

$$V^j = \sum_{k=0}^{T-1} e^{2\pi i k j/T} P_k, \qquad j = 0, \ldots, T-1, \tag{18}$$

where $P_k = P(\{2\pi k/T\})$. Since the sum in (18) is finite, it implies that for every $k \in \mathscr{Z}$

$$\frac{1}{T}\sum_{j=0}^{T-1} e^{-2\pi ikj/T}V^j = P_{\langle k\rangle}. \tag{19}$$

Let $(x(n))$ be T-PC, R be its covariance function, and let as before

$$a_j(n) := \sum_{r=0}^{T-1} e^{-2\pi ijr/T}R_x(n+r,r), \quad j \in \mathscr{Z}. \tag{20}$$

Furthermore let $\mathscr{K}$, U, V and x satisfy the conditions of Theorem 3.2. As in the case of stationary sequences the representation (13) yields the existence and a description of the spectra of T-PC sequences. This follows from the following simple observation.

Lemma 3.1 *Under the above notations*

$$a_j(m) = (U^m x, V^j x), \tag{21}$$
$$R(m+r,r) = (U^m x, P_{\langle r\rangle}x), \tag{22}$$
$$x(n) = \int_0^{2\pi} e^{-iu}G(du), \tag{23}$$

where G is a $\mathscr{K}$-valued measure given by $G(\Delta) = \dfrac{1}{T}\displaystyle\sum_{j=0}^{T-1} E(\Delta - 2\pi j/T)V^j x$.

Proof Relation (21) has been already proved in Proposition 3.1, just substitute $p = 0$ and $n = 0$ in formula (16). The relation (22) now follows from (21) and (19)

$$R(m+r,r) = \frac{1}{T}\sum_{j=0}^{T-1} e^{2\pi ijr/T}a_j(m) = \frac{1}{T}\left(U^m x, \sum_{j=0}^{T-1} e^{-2\pi ijr/T}V^j x\right) = (U^m x, P_{\langle r\rangle}x).$$

To prove (23) note that the commutativity condition $U^n V^j = e^{2\pi inj/T}V^j U^n$ together with (19) and (17) imply that

$$x(n) = P_0 U^n x = \int_0^{2\pi} e^{-iun}P_0 E(du)x, \quad n \in \mathscr{Z}.$$

Hence (23) holds true with

$$G(\Delta) = P_0 E(\Delta)x = \frac{1}{T}\sum_{j=0}^{T-1} V^j E(\Delta)x = \frac{1}{T}\sum_{j=0}^{T-1} E(\Delta - 2\pi j/T)V^j x$$

The last equality above follows from the commutation property $U^n V^j = e^{2\pi i jn/T} U^n V^j$ and (17), which imply that

$$\int_0^{2\pi} e^{-inu} E(du) V^j = V^j \left(\int_0^{2\pi} e^{-in(u-2\pi j/T)} E(du) \right) = \int_0^{2\pi} e^{-inu} V^j E(du + 2\pi j/T),$$

and hence that $V^j E(\Delta) = E(\Delta - 2\pi j/T) V^j$. Obviously G is an $\mathscr{K}$-valued measure, since E is so. In general G is not orthogonally scattered, but many values of G are so. In particular, if Δ and D are such that $(\Delta - 2\pi j/T) \cap (D - 2\pi k/T) = \emptyset$ for all $j, k \in \mathscr{Z}$, then $(G(\Delta), G(D)) = 0$. Recall that subtraction above is modulo 2π. $\square$

Substituting (17) into (21), we obtain that

$$a_j(n) = (U^n x, V^j x) = \int_0^{2\pi} e^{-iun} (E(du)x, V^j x).$$

Denoting $\gamma_j(\Delta) = (E(\Delta)x, V^j x)$, we just gave yet another proof of the following well known theorem which defines the spectrum of a PC sequence.

Corollary 3.1 (Gladyshev 1961) *Suppose that $(x(n))$ is T-PC, R is its correlation function, and $a_j(n)$ are defined by (20). Then there exist a family of measures γ_j, $j = 0, \ldots, T - 1$, such that for every $n \in \mathscr{Z}$*

$$a_j(n) = \int_0^{2\pi} e^{-imt} \gamma_j(dt). \tag{24}$$

Measures γ_j, $j = 0, \ldots, T - 1$, are referred to as the *spectral measures* of the T-PC sequence $(x(n))$.

An essential fact in the prediction theory of stationary sequences is the observation that if $h \in L^2(\mu, \mathscr{C})$ is such that $\frac{d\gamma}{d\mu}(t) = |h(t)|^2$, μ−a.e., then the sequence $(f(n))$ of functions in $L^2(\mu, \mathscr{C})$ defined as $f(n)(u) = e^{-inu} h(u)$, $n \in \mathscr{Z}$ is unitary equivalent to $(x(n))$. Recently an analogues theorem has been proved by the authors for PC sequences. Obviously one has to start with defining of a "suqare root" of the family γ_j.

Theorem 3.3 (Makagon and Miamee 2013, **Theorems 3.3 and 3.4**) *(I.) A family of measures γ_j, $j = 0, \ldots, T - 1$, is a spectrum of a T-PC sequence if and only if there exist a $2\pi/T$-invariant non-negative measure μ and a function $h \in L^2(\mu, \mathscr{C}^T)$ such that for every $j \in \mathscr{Z}$*

$$\frac{d\gamma_j}{d\mu}(t) = (1/T) h(t) h(t + 2\pi j/T)^*, \qquad \mu - e.a. \tag{25}$$

(II.) Let $(x(m))$ be a T-PC sequence, γ_j, $j = 0, \ldots, T - 1$, be its spectral measures, and μ be a $2\pi/T$ invariant measure on $[0, 2\pi)$. Suppose that $h \in L^2(\mu, \mathscr{C}^T)$

satisfies (25) above. Then the $L^2(\mu, \mathscr{C}^T)$ valued sequence $(f(m))$ defined by

$$f(m)(u) = \frac{1}{T} \sum_{j=0}^{T-1} e^{-im(u+2\pi j/T)} h(u + 2\pi j/T), \quad u \in [0, 2\pi), \ m \in \mathscr{Z}. \quad (26)$$

is unitary equivalent to the sequence $(x(m))$.

Moreover, if all γ_j, $j = 0, \ldots, T - 1$, are absolutely continuous with respect to the Lebesgue measure, then the measure μ in parts I and II can be chosen to be the Lebesgue measure.

We want to emphasize that h is $\mathscr{C}^T$ valued, so in fact comprises of T scalar functions. Moreover, neither μ nor h above are unique. The proof of this theorem is too long to be stated here.

Corollary 3.2 *Suppose that $(x(n))$ is T-PC. Then there exist a $2\pi/T$-invariant measure μ on $[0, 2\pi)$ such that the pair (U, V) in representation (13) is unitary equivalent (in the sense explained in Proposition 3.1) to the pair (M, T) of operators in some subspace of $L^2(\mu, \mathscr{C}^T)$, where $Mg(u) = e^{-iu}g(u)$ and $Tg(u) = g(u + 2\pi/T)$, i.e. M is multiplication by e^{-iu} and T is translation by $2\pi/T$.*

Proof From Theorem 3.3, II, it follows that $f(n) = \frac{1}{T} \sum_{j=0}^{T-1} e^{-2\pi inj/T} M^n T^j h$ and that $(f(n))$ and $(x(n))$ are unitary equivalent. Hence from Proposition 3.1 we conclude that the pairs (U, V) and (M, T) are unitary equivalent if restricted them to cyclic subspaces generated by x and h, respectively. $\qquad\square$

The representation (13) give rise to a T-dimensional sequence $X(n) = [X^k(n)]$ defined as

$$X^{k+1}(n) = U^n V^k x, \quad n \in \mathscr{Z}. \quad (27)$$

The T-dimensional sequence $X(n) = [X^k(n)]$ defined above is called the *induced sequence*. Proposition 3.1 shows that any two sequences induced by the same T-PC sequence $(x(n))$ are unitary equivalent. The induced sequence shares prediction properties of $(x(n))$. Additionally, the relations between spectral measures of a T-PC sequence $(x(n))$ and the spectrum of its induced sequence is quite simple. The induced sequence was introduced and thoroughly studied in Makagon (2011). Here are some properties of the induced sequence.

Proposition 3.2 *Let $(x(n))$ be T-PC, and let $(X(n))$ be the sequence induced by $(x(n))$. Then:*

1. *$(X(n))$, $n \in \mathscr{Z}$, is a T-dimensional stationary sequence.*
2. *$R^{k+1,j+1}(m, n) = (X^{k+1}(m), X^{j+1}(n)) = e^{2\pi i(m-n)k} a_{j-k}(m - n)$, $k, j = 0, \ldots, T - 1$.*
3. *$\Gamma^{k+1,j+1}(\Delta) = \gamma_{\langle j-k \rangle}(\Delta + 2\pi k/T)$, $k, j = 0, \ldots, T - 1$.*
4. *$(x(n))$ is regular iff $(X(n))$ is regular.*
5. *Rank of $(x(n))$ is equal to the rank of $(X(n))$.*

Although proofs can be extracted from Makagon (2011), we sketch them below.

Proof From the definition and relations (14) and (21) we see that

$$(X^{k+1}(m), X^{j+1}(n)) = (U^{m-n}V^k x, V^j x) = e^{2\pi i(m-n)k} a_{j-k}(m-n),$$

which gives 2. and also shows that $X(n)$ is stationary. Substituting $n = 0$ and using (24) we obtain that

$$(X^{k+1}(m), X^{j+1}(0)) = e^{2\pi imk} a_{j-k}(m) = \int_0^{2\pi} e^{-ims} \gamma_{j-k}(ds + 2\pi k).$$

Since, on the other hand, $(X^{k+1}(m), X^{j+1}(0)) = \int_0^{2\pi} e^{-ims} F^{k+1,j+1}(ds)$, this proves 3. If $\mathcal{K}$, U, V, and $x = (x(0), x(1), \ldots, x(T-1))$ are as in the proof of Theorem 3.2, then we obtain that

$$\frac{1}{T}\sum_{j=0}^{T-1} e^{-2\pi i(r+n)j/T} X^{j+1}(n) = \frac{1}{T}\sum_{j=0}^{T-1} e^{-2\pi i(r+n)j/T} U^n V^j x = \left(\frac{1}{T}\sum_{j=0}^{T-1} e^{-2\pi irj/T} V^j\right) U^n x$$

$$= P_{\langle r\rangle}(x(n), x(n+1), \ldots, x(n+T-1)) = (0, \ldots, 0, x(n+r), 0, \ldots, 0).$$

This shows that $M_X(n) = \overline{\mathrm{sp}}\{X^{j+1}(m) : m \leq n, k = 0, \ldots, T-1\} = M_x(n) \oplus M_x(n+1) \oplus \cdots \oplus M_x(n+1)$ and hence 4. To see 5. note that the space $M_X(n+1) \ominus M_X(n)$ consists of vectors $(x_1, x_2, \ldots, x_T)$ in $M_x^{\oplus T}$ such that $x_k \in M_x(n+k) \ominus M_x(n+k-1), k = 1, \ldots, T$. Since each $M_x(n+k) \ominus M_x(n+k-1)$ is of dimension at most 1, the dimension of $M_X(n+1) \ominus M_X(n)$ is T iff all $M_x(n+k) \ominus M_x(n+k-1)$, $k = 1, \ldots, T$, are nonzero. $\square$

4 An Application to the 3rd Prediction Problem

In this section we apply the theory outlined above and a known result of Wiener and Masani (1958) to obtain a certain sufficient condition for a PC sequence to be a basis for M_x. Let $(x(n))$ be a T-PC sequence and γ_j, $j = 0, \ldots, T-1$, be its spectral measures. Suppose that all γ_j's are absolutely continuous. Let γ_j' denote the density of γ_j, $\gamma'(t) = \frac{d\gamma_j}{dt}(t)$, and let $\Gamma'(t)$ be the $T \times T$ matrix whose $k+1, j+1$ entry is equal to $[\Gamma'(t)]^{k+1,j+1} = \gamma'_{\langle j-k\rangle}(t + 2\pi k/T), k, j = 0, \ldots, T-1$, that is

$$\Gamma'(t) = \begin{bmatrix} \gamma'_0(t) & \gamma'_1(t) & \gamma'_2(t) & \cdots & \gamma'_{T-1}(t) \\ \gamma'_{T-1}(t + \frac{2\pi}{T}) & \gamma'_0(t + \frac{2\pi}{T}) & \gamma'_1(t + \frac{2\pi}{T}) & \cdots & \gamma'_{T-2}(t + \frac{2\pi}{T}) \\ \gamma'_{T-2}(t + \frac{4\pi}{T}) & \gamma'_{T-1}(t + \frac{4\pi}{T}) & \gamma'_0(t + \frac{4\pi}{T}) & \cdots & \gamma'_{T-3}(t + \frac{4\pi}{T}) \\ \vdots & \vdots & \vdots & \vdots & \vdots \\ \gamma'_1(t + \frac{2(T-1)\pi}{T}) & \gamma'_2(t + \frac{2(T-1)\pi}{T}) & \gamma'_3(t + \frac{2(T-1)\pi}{T}) & \cdots & \gamma'_0(t + \frac{2(T-1)\pi}{T}) \end{bmatrix}$$

Theorem 4.1 *Suppose that $(x(n))$ is a T-PC sequence and that all spectral measures γ_j, $j = 0, \ldots, T - 1$, of $(x(n))$ are absolutely continuous. Let γ'_j be the density of γ_j, and let $\Gamma'(t)$ be the $T \times T$ matrix defined above. Let $\rho(t)$ and $\lambda(t)$ be the smallest and the largest eigenvalues of $\Gamma'(t)$. If there are numbers $\rho > 0$ and $\lambda < \infty$ such that for dt-almost all $t \in [0, 2\pi)$*

$$0 < \rho \leq \rho(t) \quad and \quad \lambda(t) \leq \lambda < \infty, \tag{28}$$

then $(x(n))$ is a basis for M_x.

Proof Let $X(n)$ be the sequence induced by $(x(n))$. From Proposition 3.2 it follows that $\Gamma'(t)$ is the density of the spectral measure of $X(n)$. From Wiener and Masani (1958), Sect. 5, we conclude that the sequence $(z(n))$ made form $X(n)$ by listing all elements $X^k(n), k = 1, \ldots, T, n \in \mathscr{Z}$, of $X(n)$ in a linear order, i.e. $z(nT + k - 1) = X^k(n)$, form a basis for M_X. Note that, by (15), $x(k) = \frac{1}{T} \sum_{j=0}^{T-1} e^{-2\pi i k j/T} z(kT + j)$, and hence

$$\sum_{k=p}^{q} a_k x(k) = \sum_{k=p}^{q} \sum_{j=0}^{T-1} \left(\frac{1}{T} a_k e^{-2\pi i k j/T} \right) z(kT + j) = \sum_{l=pT}^{qT+T-1} c_l z(l). \tag{29}$$

for properly defined c_l's. Since $(z(n))$ is a basis, by Proposition 2.2 there exists a universal constant K such that for every $-\infty < p \leq m \leq n \leq q < \infty$.

$$\left\| \sum_{l=mT}^{nT+T-1} c_l z(l) \right\| \leq K \left\| \sum_{l=pT}^{qT+T-1} c_l z(l) \right\|.$$

This and (29) above imply that for any choice of scalars (a_n) and for any integers $-\infty < p \leq m \leq n \leq q < \infty$

$$\left\| \sum_{k=m}^{n} a_k x(k) \right\| \leq K \left\| \sum_{k=p}^{q} a_k x(k) \right\|.$$

Hence $(x(n))$ is a basis for M_x. $\qquad\square$

Since (ba) $\Rightarrow$ (ar), an immediate corollary of Theorem 4.1 is that

Corollary 4.1 *Suppose that $(x(n))$ is a T-PC sequence satisfying the assumptions of Theorem 4.1. Then for every $n \in \mathscr{Z}$ there exists a unique sequence $\phi_k(n)$, $k = 1, 2, \ldots$ such that*

$$(x(n)|M_x(n-1)) = \sum_{k=1}^{\infty} \phi_k(n)x(n-k).$$

Moreover, for every $k = 1, 2, \ldots$ the sequence $\phi_k(n)$ is T-periodic in $n \in \mathscr{Z}$.

The moreover part follows from the uniqueness of the coefficients $\phi_k(n)$ and from the fact that if W is the T-shift of $(x(n))$, then $W(x(n)|M_x(n-1)) = (x(n+T)|M_x(n+T-1))$.

Acknowledgments This chapter deals with the theory of PC sequences, which is a very small part of a broad and multidisciplinary area of analysis of periodically correlated processes and cyclostationary signals. No paper on PCs would be complete without mentioning names as Hurd, Yavorskij, Leśkow, Dehay, Neapolitano, Weron, Wylomanska, who have shaped the theory and practice of periodically correlated processes (cf. Broszkiewicz-Suwaj et al. 2004; Cambanis et al. 1994; Dehay 1994; Dragan and Yavorskii 1985; Hurd 1974, 1989, 1991; Hurd and Leskow 1992a, b; Hurd et al. 2002; Hurd and Miamee 2007; Javors'kyj et al. 2003, 2007, 2010; Lenart et al. 2008; Leskow and Weron 1992; Neapolitano 2012; Weron and Wylomanska 2004; Wylomanska 2008) and provided a motivation for our study. We are grateful to Professor Leśkow for organizing annual workshops which give this diverse community an opportunity to share their research, experience, and ideas.

References

Babenko KI (1948) On conjugate functions. Dokl Akad Nauk SSSR 62:157–160

Broszkiewicz-Suwaj E, Makagon A, Weron R, Wylomanska A (2004) On detecting and modeling periodic correlation in financial data. Phys A 336(1–2):196–205

Cambanis S, Houdr C, Hurd H, Leskow J (1994) Laws of large numbers for periodically and almost periodically correlated processes. Stochastic Process Appl 53(1):37–54

Dehay D (1994) Spectral analysis of the covariance of the almost periodically correlated processes. Stochastic Process Appl 50(2):315–330

Dragan Ya P, Yavorskii IN (1985) Statistical analysis of periodic random processes (Russian). Otbor i Peredacha Informatsii 71:20–29

Folland GB (1995) A course in abstract harmonic analysis. Studies in advanced mathematics. CRC Press, Boca Raton

Gladyshev EG (1961) Periodically correlated random sequences. Soviet Math 2:385–388

Heil C (2011) A basis theory primer. Birkhäuser, Boston

Helson H (1964) Lectures on invariant subspaces. Academic Press, New York

Helson H, Szegö G (1960) A problem in prediction theory. Ann Mat Pura Appl 51(4):107–138

Hurd HL (1974) Stationarizing properties of random shifts. SIAM J Appl Math 26:203–212

Hurd HL (1989) Representation of strongly harmonizable periodically correlated processes and their covariances. J Multivariate Anal 29(1):53–67

Hurd HL (1991) Correlation theory of almost periodically correlated processes. J Multivariate Anal 37(1):24–45

Hurd HL, Leskow J (1992) Strongly consistent and asymptotically normal estimation of the covariance for almost periodically correlated processes. Statist Decisions 10(3):201–225

Hurd HL, Leskow J (1992) Estimation of the Fourier coefficient functions and their spectral densities for α-mixing almost periodically correlated processes. Statist Probab Lett 14(4):299–306

Hurd H, Makagon A, Miamee AG (2002) On AR(1) models with periodic and almost periodic coefficients. Stochastic Process Appl 100:167–185

Hurd HL, Miamee AG (2007) Periodically correlated random sequences. In: Spectral theory and practice. Wiley Series in Probability and Statistics. Wiley-Interscience, New Jersey

Javors'kyj I, Mykhailyshyn V, Zabolotnyj O (2003) Least squares method for statistical analysis of polyrhythmics. (English summary). Appl Math Lett 16(8):1217–1222

Javors'kyj I, Isayev I, Zakrzewski Z, Brooks SP (2007) Coherent covariance analysis of periodically correlated random processes. Signal Process 87:13–32

Javors'kyj I, Isayev I, Majewski J, Yuzefovych R (2010) Component covariance analysis for periodically correlated random processes. Signal Process 90:1083–1102

Lenart L, Leskow J, Synowiecki R (2008) Subsampling in testing autocovariance for periodically correlated time series. J Time Ser Anal 29(6):995–1018

Leskow J, Weron A (1992) Ergodic behavior and estimation for periodically correlated processes. Statist Probab Lett 15(4):299–304

Lindenstrauss J, Tzafiri L (1997) Classical Banach spaces I. Springer, Berlin

Makagon A (1984) Interpolation error operator for Hilbert space valued stationary stochastic processes. Probab Math Statist 4(1):57–65

Makagon A (2011) Stationary sequences associated with a periodically correlated sequence. Probab Math Stat 31(2):263–283

Makagon A, Miamee AG (2013) Spectral representation of periodically correlated sequences. Probab Math Stat 33(1):175–188

Makagon A, Salehi H (1989) Notes on infinite-dimensional stationary sequences. In: Probability theory on vector spaces, IV, Lecture Notes in Mathematics 1391, Springer, Berlin, pp 200–238

Makagon A, Weron A (1976) q -variate minimal stationary processes. Studia Math 59(1):41–52

Masani P (1960) The prediction theory of multivariate stochastic processes III. Unbounded spectral densities. Acta Math 104:141–162

Masani P (1966) Recent trends in multivariate prediction theory. In: Krishnaiah PR (ed) Multivariate analysis, Proceedings of the International Symposium Dayton, Ohio 1965, Academic Press, New York, pp 351–382

Miamee AG (1993) On basicity of exponentials in $L^p(\mu)$ and general prediction problems. Period Math Hungar 26(2):115–124

Miamee AG (1991) The inclusion $L^p(\mu) \subseteq L^q(\nu)$. Amer Math Monthly 98(4):342–345

Neapolitano A (2012) Generalizations of cyclostationary signal processing. Spectral analysis and applications. IEEE Press, Chichester

Pourahmadi M (2001) Foundation of time series analysis and prediction theory. Wiley, New York

Weron A, Wylomanska A (2004) On ARMA(1, q) models with bounded and periodically correlated solutions. Probab Math Statist 24(1), Acta Univ. Wratislav. No. 2646, 165–172

Wiener N, Masani P (1957) The prediction theory of multivariate stochastic processes I. The regularity condition. Acta Math 98:111–150

Wiener N, Masani P (1958) The prediction theory of multivariate stochastic processes II. The linear predictor. Acta Math 99:93–137

Wylomanska A (2008) Spectral measures of PARMA sequences. J Time Ser Anal 29(1):1–13

Methods of Periodically Correlated Random Processes and Their Generalizations

I. Javors'kyj, R. Yuzefovych, I. Kravets and I. Matsko

Abstract The results obtained by authors in the area of theory and methods of statistical analysis of periodically correlated random processes and their generalizations are presented in this article. The main methods for estimation of their correlation and spectral characteristics: coherent, component, least square method and linear filtration method are analyzed. The ways of generalization of these methods to the case of unknown a priori period of non-stationarity are considered and the possible algorithms of its estimation are presented.

1 Introduction

Both rough recurrence and stochastic are characteristic feature of time changeability of many physical processes. Recurrence of the oscillations property—rhythmic can be caused by both the effect of external forces (forced oscillations) on a given system and results from the internal interrelations (eigen oscillations) existing in systems. Rhythmical processes are encountered in many spheres of science and engineering including radiophysics, geophysics, oceanology, meteorology, climatology, vibrodiagnostics, hydroacoustics, biology, seismology, economics, telecommunication etc.

Processes which are caused by astrophysical factors should be noted among all the diversity of forced oscillations occurring in nature. These are: the annual and diurnal oscillations of geophysical, oceanological and meteorological quantities (Dragan et al. 1987), which are the result of the earth's revolution around the sun and rotation

I. Javors'kyj · R. Yuzefovych · I. Kravets (✉) · I. Matsko
Karpenko Physico-mechanical Institute, National Academy of Sciences of Ukraine,
Lviv, Ukarine
e-mail: dr.kravets@gmail.com

I. Javors'kyj
Institute of Telecomunication, University of Technology and Life Sciencess, Bydgoszcz, Poland
e-mail: javor@utp.edu.pl

F. Chaari et al. (eds.), *Cyclostationarity: Theory and Methods*,
Lecture Notes in Mechanical Engineering, DOI: 10.1007/978-3-319-04187-2_6,
© Springer International Publishing Switzerland 2014

of earth round its axis of the equator; the tidal oscillations of the sea level, the earth's crust, sea currents, internal waves whose polyrhythmic are caused by the polyharmonic character of the potential of tideforming forces. Among the autooscillation processes we should note the oscillations in auto-generators of various physical nature (Gudzenko 1959; Malakhov 1968; Rytov 1976), signals of geomagnetic pulsations (Mikhailyshyn et al. 1990), vibrations (Mikhailyshyn et al. 1990; Antoni 2009) and oscillations in biological systems (bio-rhythmic) (Aschoff 1981).

A certain model of mathematical oscillations is the methodological base for investigation of the oscillation processes structure on the basis of experimental data. In the pioneering investigations of rhythmic, the evident advantages were given to deterministic conception which is based on models in the form of periodical functions. The aspiration to take into account random features of oscillations encouraged the development of probabilistic methods that consider phenomena as stationary random processes. Within the framework of such an approach, rhythmic features of physical processes manifest themselves in an oscillatory behavior of correlogram and existence of several peaks of the estimates of the spectral density. However, these characteristics describe the average properties of processes and do not provide information on their temporal structure, which can be determined, though in an idealized form, with use of deterministic models. Natural combination and development of these two approaches give rise to a concept that represents probabilistic models as periodically correlated random processes (PCRP) and PCRP-related processes (bi-,poly-,and almost periodically correlated). Such models generalize the notion of the recurrence to situations where stochasticity plays a significant role. These models provide an opportunity to describe the structure of rhythmic variations more thoroughly and objectively and include the above mentioned models as particular cases. The PCRP model allow us not only to analyze processes using special methods characteristic of each model, but also to investigate the different phenomena in common terms for all models.

In this chapter we consider probabilistic characteristics of periodically nonstationary probabilistic models, analyze the properties of such models and develop the general approach for the estimation of such characteristic from the empirical data. We analyze the methods of statistical estimation and present radically new results of the investigation of hidden periodicities.

The manuscript presents a survey of results of the authors investigations realized at Karpenko Physico-Mechanical Institute of National Academy of Sciences of Ukraine in Lviv and at University of Technology and Life Sciences in Bydgoszcz.

2 Periodically Correlated Random Processes and Their Generalizations as Probabilistic Model of Stochastic Oscillations

The methods of periodically correlated random processes and their generalizations extend our capabilities for understanding the regularities of stochastic oscillations. These methods provide additional opportunities associated with the description of the properties of oscillations in terms of time-varying probabilistic characteristic. Within the framework of the second order theory, the hidden periodicity of physical processes is manifested as periodic temporal variations of the mean and the correlation function:

$$m(t) = E\xi(t),$$

$$b(t, u) = E\overset{\circ}{\xi}(t)\overset{\circ}{\xi}(t + u) = b(t + T, u),$$

$$\overset{\circ}{\xi}(t) = \xi(t) - m(t). \tag{1}$$

The properties of (1) define the class of PCRP. Apparently, the first mathematically correct definition of PCRP was given by Koronkevich (1957). This definition is applicable to the description of the properties of solutions for differential equations with periodic coefficients and random right-hand sides. The adequate models of communications signals PCRP are discussed in Bennet (1958) and Franks (1969). The possibility of using such processes to describe stochastic oscillations was also mentioned by Stratonovich (1961). Several papers (Papoulis 1983; Kozel 1959; Myslovich et al. 1980; Tikhonov 1956) consider transformations of periodically nonstationary and periodically correlated signals. Correlation and spectral properties of PCRP were studied by Rytov (1976), Gladyshev (1959), Gudzenko (1959), Ogura (1971), Papoulis (1983), Gardner and Franks (1975), Gardner (1985, 1986, 1994) and Hurd (1969, 1989). Gudzenko employed PCRP to analyze fluctuations in autooscillation systems. Rytov (1976) indicated the possibility for applying PCRP to study noise in cyclic remagnetization. Romanenko and Sergeyev (1968) pointed to the adequacy of employing the PCRP methods to describe a turbulent flow of water near a ship screw propeller and to investigate temporal variations in the electric-power consumption and physiological changes. Dragan (1969, 1970, 1975, 1978, 1980) developed the fundamentals of the PCRP theory with a limited average power. In collaboration with K.S. Voichishin. Dragan also applied the PCRP model to formulate certain general properties of the stochastic model of rhythmic (Dragan 1985; Voichishyn and Dragan 1971; Dragan 1972). Voichyshyn (1975) made the first attempts for using this model to analyze daily rhythmic variations in certain geophysical processes. Dragan and Javors'kyj extended this approach to study the properties of wind waves (Dragan and Javors'kyj 1975). In both cases, the estimates of the mean and variance were analyzed. Note that, in earlier studies, the PCRP model, which was understood even in a more comprehensive sense (with use of histograms and estimates of the correlation function), was applied to investigate daily variations in the temperature and humidity of the air and the soil temperature (Zhukovsky 1969; Kiselyeva and Chudnovsky 1968; Mamontov 1968a,b; Mishchenko 1960, 1966). An analogous modification of this model was used to study diurnal and seasonal changes in meteorological processes (Zhukovsky 1969). Groisman (1977) considered the periodic correlation of precipitation series.

However, the systematic application of the PCRP model for analysis of rhythmic signals was limited mainly because of the absence of an appropriate procedure of data processing (Kolyesnikova and Monin 1965). Chapters (Dragan et al. 1987;

Mikhailyshyn et al. 1990; Javorskyj et al. 2006, 2007, 2010; Jaworskyj et al. 2011a,b; Kravets 2012) were devoted to the development of such a procedure and application of this method for the investigation of the structure of rhythmic variations in physical processes.

Within the framework of the PCRP model of rhythmic, the mean describes the regular periodical oscillations, the variance characterizes the periodicity of the power of fluctuations around this regular behavior and the correlation function describes the character of periodic changes in correlations between the values of fluctuating parameters at various moments of time separated by equal time intervals.

If we assume that the PCRP mean and the correlation function are absolutely integrable within the interval $[0, T]$, then these characteristics can be represented in Fourier series:

$$m(t) = \sum_{k \in Z} m_k e^{ik\frac{2\pi}{T}t},$$

$$b(t, u) = \sum_{k \in Z} B_k(u) e^{ik\frac{2\pi}{T}t}, \tag{2}$$

where $|m_k| \to 0$ and $|B_k(u)| \to 0$ as $k \to \infty$. The coefficients m_k and $B_k(u)$ (the latter coefficients are referred to as correlation components) quantitatively characterize the waveforms of the periodic curves representing the mean and the correlation function. The correlation components $B_k(u)$ satisfy the equations:

$$B_k(-u) = B_k(u) e^{ik\frac{2\pi}{T}u},$$

$$B_k(u) = \overline{B_{-k}(u)}. \tag{3}$$

The zero-th correlation component is an even $B_0(-u) = B_0(u)$ and positive-definite function. In other words, this component has all the properties of the correlation function of a stationary random process.

The representation of the PCRP in terms of stationary connected components (Dragan et al. 1987; Ogura 1971; Hurd 1989; Dragan 1969):

$$\xi(t) = \sum_{k \in Z} \xi_k(t) e^{ik\frac{2\pi}{T}t} \tag{4}$$

is important for understanding the structure of PCRP as a model of stochastic oscillations. As can be seen from expression (4), PCRP can be represented as a sum of amplitude and phase-modulated harmonics, whose frequencies are multiple of the fundamental oscillation frequency $\omega_0 = \frac{2\pi}{T}$. Comparing (2) and (4), we find that the components m_k coincide with the mean of stationary random processes $\xi_k(t)$. The autocorrelation functions of these processes $D_{kk}(u) = E\,\xi_k^\circ(t + u)\overline{\xi_k^\circ(t)}$ determine the PCRP zero-th correlation component:

$$B_0(u) = \sum_{k \in Z} D_{kk}(u) e^{ik\frac{2\pi}{T}u}. \tag{5}$$

The cross-correlation functions of components whose numbers are shifted by l determine the l th correlation components:

$$B_l(u) = \sum_{k \in Z} D_{k-l,k}(u)\, e^{ik\frac{2\pi}{T}u}. \tag{6}$$

As can be seen from the representation (4), the PCRP model covers various simpler models of rhythmic variations, including the additive model $\xi(t) = \eta(t) + f(t)$, where $\eta(t)$ is a stationary random process and $f(t)$ is a periodic function; the multiplicative model $\xi(t) = \eta(t)\, f(t)$; and the additive-multiplicative model $\xi(t) = g(t) + \eta(t)\, f(t)$, where $g(t)$ is a periodic function. The first model is often used in hydro-meteorological studies. The second and third models are mainly applied to describe variations in the variance, that is, in the power of fluctuation oscillations (Gruza 1982; Poljak 1978).

The PCRP variable spectral density $f(\omega, t)$ is a complex function: $f(\omega, t) = \mathrm{Re}\, f(\omega, t) - i\, \mathrm{Im}\, f(\omega, t)$. Its real part is determined by the cosine transform of the even part of the correlation function:

$$\mathrm{Re}\, f(\omega, t) = \frac{1}{2\pi} \int_0^\infty b^s(t, u) \cos \omega u\, du, \tag{7}$$

whereas the imaginary part is given by the sine transform of the odd part of the correlation function:

$$\mathrm{Im}\, f(\omega, t) = \frac{1}{2\pi} \int_0^\infty b^c(t, u) \sin \omega u\, du, \tag{8}$$

where $b(t, u) = b^s(t, u) + b^c(t, u)$, $b^s(t, u) = -b^s(t, -u)$ and $b^c(t, u) = b^c(t, -u)$.

The real and imaginary parts of the spectral density are even and odd functions of the frequency, respectively:

$$\mathrm{Re}\, f(-\omega, t) = \mathrm{Re}\, f(\omega, t),$$
$$\mathrm{Im}\, f(\omega, t) = -\mathrm{Im}\, f(-\omega, t).$$

Since

$$b(t, u) = 2 \int_0^\infty [\mathrm{Re}\, f(\omega, t) \cos \omega u - \mathrm{Im}\, f(\omega, t) \sin \omega u]\, du,$$

for $u = 0$ we have:

$$b\left(t,0\right) = 2 \int_{0}^{\infty} \mathrm{Re}\, f\left(\omega,t\right) d\omega. \tag{9}$$

This formula allows us to provide a physical interpretation of the function $\mathrm{Re}\, f\left(\omega,t\right)$. Since $b\left(t,0\right)$ characterizes the instantaneous power of the process. $\mathrm{Re}\, f\left(\omega,t\right)$ describes the distribution of this power in the $\left(\omega,t\right)$ plane. The integration of this quantity with respect to all the relevant frequencies yields the value of the power at a given moment of time t. However, we cannot interpret this quantity as the power spectral density because this quantity is not necessarily non-negative for all $\left(\omega,t\right)$, although, similarly to the spectral density of a stationary random process, this function is even. For $b\left(t,0\right) = const$, expression (9) is reduced to the well-known relation for stationary random processes. Then $\mathrm{Re}\, f\left(\omega,t\right) = f\left(\omega\right)$ is the power spectral density. This function can be interpreted in terms of energy characteristics for the so-called quasi-stationary random process, when the rate of the variation of the correlation function, due to the lag is much higher than the rate of temporal variation of this correlation function. Such a quasi-stationary behavior may also occur in the case of PCRP. Then, we have $\mathrm{Re}\, f\left(\omega,t\right) \geq 0$, and we can consider this function as the power spectral density. Generally, we cannot use this physical interpretation. As can be seen from the formula $b^{c}\left(t,u\right) = \frac{1}{2}\left[b\left(t,u\right) - b\left(t-u,u\right)\right]$, if the correlation function rapidly decays with the growth in the lag and displays only small variations in the argument t within the interval $\left[t-u,t\right]$, then $b^{c}\left(t,u\right)$ is small. Hence, the function $b^{c}\left(t,u\right)$ can be used to characterize transient processes, and $\mathrm{Im}\, f\left(\omega,t\right)$ describes the properties of such processes in the frequency domain.

The variable spectral density $f\left(\omega,t\right)$ is a periodic function of time. The amplitudes of harmonics of this function determine the spectral component $f_k\left(\omega\right)$:

$$f\left(\omega,t\right) = \sum_{k \in Z} f_k\left(\omega\right) e^{ik\frac{2\pi}{T}}.$$

These components are defined by means of the Fourier transform of the correlation components:

$$f_k\left(\omega\right) = \frac{1}{2\pi} \int_{0}^{\infty} B_k\left(u\right) e^{-i\omega u} du. \tag{10}$$

If the correlation components are absolutely integrable, then $f_k\left(\omega\right)$ is continuous for all $\omega \in R$ and $\left|f_k\left(\omega\right)\right| \to 0$ as $\omega \to \pm\infty$.

Using expressions (3) we derive:

$$f_{-k}\left(\omega\right) = \overline{f_k\left(-\omega\right)},$$

$$f_k\left(-\omega\right) = f_k\left(\omega + k\frac{2\pi}{T}\right).$$

The zero-th spectral component $f_0(\omega)$ is a real even function and $f_0(\omega) \geq 0$ for all $\omega \in R$. This conclusion is rather natural because $f_0(\omega)$ is the Fourier transform of the zero-th correlation component $B_0(u)$, which coincides with the correlation function corresponding to the stationary approximation of PCRP. The zero-th spectral component describes the frequency distribution of the average power of oscillations, whereas higher order spectral components characterize frequency properties of cross-correlations $\xi_k(t)$ of modulating processes:

$$f_k(\omega) = \sum_{l \in Z} f_{l-k,l}(\omega - l\omega_0).$$

Vector PCRP are natural probabilistic models of rhythmic variations of vector physical quantities. The mean of vector PCRP are described by periodic vectors $m_{\overline{v}}(t) = m_{\overline{v}}(t+T)$. whereas correlation functions $d_{\overline{v}}(t, u)$ and spectral densities $f_{\overline{v}}(\omega, t)$ are given by periodic dyad tensors. Similarly to vector stationary random processes (Dragan et al. 1987; Bjelyshev et al. 1983), the properties of vector PCRP can be described in terms of the invariants of tensors $b_{\overline{v}}(t, u)$ and $f_{\overline{v}}(\omega, t)$ (Javors'kyj 1987). These invariants unambiguously characterize the correlation and spectral structure of vector random processes regardless of the choice of the coordinate frame.

More sophisticated models of rhythmic variations are based on bi-and poly-PCRP (Dragan 1972) which describe both the interference and nonlinear interaction of oscillations with different periods. To analyze modulation effects in the PCRP model, we should represent a PCRP as a series (4). Then, the interaction of two rhythms gives rise to a process:

$$\xi(t) = \sum_{j,k \in Z} \xi_{k,j}(t) e^{i\Lambda_{kj}t},$$

where $\Lambda_{kj} = k\frac{2\pi}{T_1} + j\frac{2\pi}{T_2}$. The mean and the correlation function of such process, which is referred to as the bi-PCRP, are written as:

$$m(t) = \sum_{j,k \in Z} m_{kj} e^{i\Lambda_{kj}t}, b(t, u) = \sum_{j,k \in Z} m_{kj}(u)^{i\Lambda_{kj}t}.$$

The mixed components of the mean m_{kj} and the correlation function $B_{kj}(u)$ characterize the modulation interaction of the kth and jth harmonics. The structure of matrices representing these quantities is indicative of the presence of bi-rythmic variations of a certain type.

Generalizing the model of bi-PCRP to additive-multiplicative interaction involving many rhythms, we arrive at the notion of poly-PCRP, which can be represented in the following form:

$$\xi(t) = \sum_{l_1,\,\ldots,\,l_N \in Z} \xi_{l_1,\,\ldots,\,l_N}(t)\, e^{\,it \sum_{j=1}^{N} l_j \Lambda_j}.$$

Components $\xi_{l_1,\ldots,l_N}(t)$ characterize modulation interactions of N rhythms with periods T_j. Poly-PCRP, in their turn, form a subclass of almost-PCRP. which can be represented as

$$\xi(t) = \sum_{j \in Z} \xi_j(t)\, e^{i\omega_j t},$$

where $\xi_j(t)$ are the stationary connected processes. Almost-PCRP are reduced to poly-PCRP if the Fourier exponents ω_j, can be written as:

$$\omega_j = \sum_{k=1}^{N} r_{jk} \Lambda_k, \tag{11}$$

where r_{jk} are integers. In this case, the basis $\left\{ \Lambda_k, k = \overline{1, N} \right\}$ is referred to as finite and integer basis of the set $M = \{\omega_k, k \in Z\}$. If the set M is an arithmetic progression, i.e., $\omega_j = j\omega_0$, $\omega_0 = const$ is the difference of this progression, then the class of almost-PCRP is reduced to the class of PCRP. We can describe the latter class using the invariance of its characteristics with respect to shifts by $T = 2\pi/\omega_0$.

In contrast to the subclasses that include poly-and bi-PCRP models, the general model of rhythmic oscillations in the form of almost-PCRP allow us to investigate only the interference of rhythms with frequencies ω_j. We can reveal nonlinear properties of the relevant processes and phenomena by studying the structure of elementary oscillations in greater detail with the use of relationship (11), which is satisfied for bi- and poly-PCRP models.

3 Estimation of Probabilistic Characteristics of Oscillations

As can be seen from the aforesaid, the requirements to the model of rhythmic variations in the form of PCRP and their generalizations are formulated in terms of the type of temporal variations of probabilistic characteristics. Importantly, these requirements are also applicable to the second-order moment functions. Within the framework of such an approach, we should abandon the assumption that rhythmic variations are reduced to a visible recurrence of values, and perturbations distort a strictly periodic pattern. This assumption provides the basis for a wide use of the additive model for the description of seasonal rhythmic variations of geophysical processes. However, processes that occur in the atmosphere and ocean are essentially nonlinear. Therefore, it is doubtful whether we could represent the temporal variations of the quantities under consideration as mutually independent regular seasonal changes and random fluctuations. The PCRP model is based on the assumption that the relevant processes are interdependent. Obviously, the substantiation of the concept related to this model should be based on empirical data, and we should examine PCRP methods of statistical analysis adequate to the problem under study. Initial approaches to this problem were indicated by Gudzenko (1959, 1961), who demonstrated that it is possible to estimate the mean and the correlation function either by evaluating appropriate Fourier components or by analyzing the counts sampled through a time interval equal to the correlation period. The first method is referred to as component and the second is called coherent. Dragan (1972) thoroughly investigated the properties of PCRP sequences of counts. Based on the periodicity on the average, Dragan also substantiated the applicability of such sequences for statistical estimation of PCRP characteristics (Dragan 1972). Subsequent stages in the development of the methods of PCRP statistics (Dragan et al. 1987; Mikhailyshyn and Javors'kyj 1990; Javorskyj et al. 2007, 2011, 2012; Yavors'kyi et al. 2009) involved the solution of problems of estimation theory in the context of the systematic development of the means for statistical analysis of empirical data.

Evidently, the coherent and component methods of estimation can be deduced from the periodicity of the probabilistic characteristics of PCRP. Coherent estimates for the mean and the correlation function are written as:

$$\widehat{m}\,(t) = \frac{1}{N} \sum_{n=0}^{N-1} \xi\,(t + nT)\,,$$

$$\widehat{b}\,(t, u) = \frac{1}{N} \sum_{n=0}^{N-1} \left[\xi\,(t + u + nT) - \widehat{m}\,(t + u + nT)\right]\left[\xi\,(t + nT) - \widehat{m}\,(t + nT)\right].$$

Component estimates for the mean and the correlation function are based on the estimates for the components of the relevant Fourier series, i.e.:

$$\widehat{m}\,(t) = \sum_{l=-N_1}^{N_1} \widehat{m}_l e^{il\frac{2\pi}{T}t}\,, \tag{12}$$

$$\widehat{b}\,(t, u) = \sum_{l=-N_2}^{N_2} \widehat{B}_l\,(u)\, e^{il\frac{2\pi}{T}t}\,, \tag{13}$$

where:

$$\widehat{m}_l = \frac{1}{\theta} \int_0^\theta \xi\,(t)\, e^{-il\frac{2\pi}{T}t} dt\,, \tag{14}$$

$$\widehat{B}_l\,(u) = \frac{1}{\theta} \int_0^\theta \left[\xi\,(t) - \widehat{m}\,(t)\right]\left[\xi\,(t + u) - \widehat{m}\,(t + u)\right] e^{-il\frac{2\pi}{T}t} dt\,. \tag{15}$$

In formulas (12) and (13), N_1 and N_2 are the numbers of components estimated for the mean and the correlation function, respectively. The estimates for the mean and the correlation function of PCRP based on coherent averaging of the counts sampled with a time interval equal to the correlation period T employ only one value of the process within the period. Statistics (12) and (13) are based on all the values of the relevant continuous realization. Therefore, if the correlation function of a PCRP displays considerable changes within the interval equal to the correlation period, then, for a given the realization length, component estimates are characterized by a smaller variance that coherent estimates. This advantage of component estimates is especially important if the number of components of characteristics being evaluated is small because the variance of the component estimate increases with the growth of this number.

Coherent and component methods of estimation of probabilistic PCRP characteristics can be considered as particular cases of a more general method of estimation— the method of linear filtration (Dragan et al. 1987; Javors'kyj 1987; Yavorskyj et al. 2012):

$$\widehat{m}(t) = \int_0^\theta \xi(t - \tau) h(\tau) \, d\tau, \tag{16}$$

$$\widehat{b}(t, u) = \int_0^\theta \overset{\circ}{\xi}(t + u - \tau) \overset{\circ}{\xi}(t - \tau) h(\tau) \, d\tau. \tag{17}$$

If the condition:

$$\int_0^\theta h(\tau) e^{-ik\frac{2\pi}{T}\tau} d\tau = 1, \, k = -\overline{N_1, N_1} \tag{18}$$

is satisfied, then $E\widehat{m}(t) = m(t)$, i.e., (16) provides an unbiased estimate. Suppose that $h(\tau)$ is a periodic function, $h(\tau + T) = h(\tau)$. Then, we can write:

$$h(\tau) = \sum_{i \in Z} h_l e^{i\frac{2\pi}{T}\tau}.$$

Hence, with allowance for (17) with $\theta = NT$, we find that $h_l = \theta^{-1}$. Expressions (16) and (17) in this case are transformed into the formulae for coherent estimates. If the number of harmonics is finite, the function $h(\tau)$ describes the impulse response a component filter. Thus, the class of estimates described by (16) and (17) with a periodic weight function $h(\tau)$ is completely covered with coherent and component estimates.

The frequency characteristics of the coherent and component comb filters are given by:

$$H\left(\omega\right) = e^{-i\frac{\omega T}{2}(N-1)} \sin \omega \frac{N}{2} T \left(2\pi N \sin \frac{\omega T}{2}\right)^{-1}$$

and

$$H\left(\omega\right) = \frac{1}{2\pi} \sum_{k=-M}^{M} e^{-i\left(\omega-k\frac{2\pi}{T}\right)\frac{\theta}{2}} \sin\left(\omega - k\frac{2\pi}{T}\right)\frac{\theta}{2}\left[\left(\omega - k\frac{2\pi}{T}\right)\frac{\theta}{2}\right]^{-1},$$

respectively. The component filter is characterized by a lower level of side lobes as compared with the coherent filter and by a finite number of transmission bands. The number of these bands is determined by the number M of components being estimated. The above-specified properties of the filters mainly account for a higher reliability of component estimates. The difference between component and coherent estimates vanishes as $M \to \infty$.

Component estimates are parametric estimates: they are based on the unknown parameters $\hat{m}_k$ and $\hat{B}_k\left(u\right)$ of trigonometric polynomial. We may use least squares technique for their estimation (Javors'kyj 1988; Mikhailyshyn and Javors'kyj 1990; Zabolotnyj et al. 2000; Yavorskyj et al. 2011). The least squares estimates are estimated by minimizing the functionals:

$$F_1\left(\hat{m}_0, \hat{m}_1^c, \ldots, \hat{m}_{N_1}^s\right) = \int_0^\theta \left[\xi\left(t\right) - \hat{m}\left(t\right)\right]^2 dt,$$

$$F_2\left(\hat{B}_0\left(u\right), \hat{B}_1^c\left(u\right), \ldots, \hat{B}_{N_1}^s\left(u\right)\right) = \int_0^\theta \left[\overset{\circ}{\xi}\left(t\right)\overset{\circ}{\xi}\left(t+u\right) - \hat{b}\left(t, u\right)\right]^2 dt,$$

where $\hat{m}\left(t\right)$ and $\hat{b}\left(t, u\right)$ are obtained from (12) and (13), here $\hat{m}_k = \frac{1}{2}\left[\hat{m}_k^c - i\hat{m}_k^s\right]$, $\hat{B}_k\left(u\right) = \frac{1}{2}\left[\hat{B}_k^c\left(u\right) - i\hat{B}_k^s\left(u\right)\right]$. The bias of least squares estimates does not depend on the realization length θ. And under condition $\lim_{u\to\infty} b\left(t, u\right) = 0$ are convergent. If the realization length $\theta = NT$ the component and least squares estimates match.

The better quality of component and least squares estimates is due to the use of apriori data concerning the number of components of characteristics being estimated. One can employ apriori data concerning the correlation structure of the PCRP to find more efficient estimates. One of the possible ways to improve the estimation efficiency is to choose optimal linear filter (Mezentsev and Javors'kyj 1988; Yavorskyj et al. 2012). In particular, the minimum, on the average, variance of estimates can be achieved with the use of a linear invariant filter, and the minimum variance for an arbitrary moment of time is achieved by means of a filter with periodically varying parameters.

For a detail analysis of PCRP structure the estimation method based on the stationary components extraction (Yavorskyj et al. 2011) should be used. These methods use the harmonic series representation of PC process, that is the generalization of

Fourier series for periodic functions in the sense that Fourier coefficients are replaced by jointly stationary processes. In order to extract stationary components the spectral region is divided into the bands $[(k - 0.5)\,\omega_0, (k + 0.5)\,\omega_0]$, $k \in Z$. First method consists in shifting of each band on $k\omega_0$ value and low-band filtration in the interval $[-\omega_0/2, \omega_0/2]$. The second method filtrates each part and uses the Hilbert transform for components estimation. Probability characteristics of these components are determined by the probability characteristics of PC process. They are jointly stationary random processes, which spectral density functions are located in the interval $[-\omega_0/2, \omega_0/2]$, that is extracted stationary components have finite spectrum. If PC process is narrow band one, namely it is formed by harmonic components, which are modulated by low band stationary random processes, then extracted components are very close to these ones which form the PC process. If modulated stationary processes are wide band processes, then extracted components will have characteristics, which are formed as result of shifting, filtration and superposition of first ones. Note, that PC processes harmonic series representation through stationary components with finite or infinite spectra is equivalent in the meaning of PC process probability characteristics. It allows us to use combing filtering method for PC process probability characteristics calculation. Developed method deals with the structure of a process (we know which stationary component pairs have nonzero correlations and may easily build the model of the process) as opposed to coherent, component and least squares methods which deal only with integral characteristics of the process. Such approach simplifies the properties parameterization of estimated correlation function and is very effective for vibration diagnostics (Bjelyshev et al. 1983; Gudzenko 1961), because of the probabilities of extracted stationary components are directly connected to the faults of mechanical rotary systems. Also this idea can be put as a base for PC process parametric modeling development (Kravets 2012). Properly speaking, the mixture of the harmonic series representation and vector autoregressive moving average parametric model is the alternative method to periodic autoregressive moving average parametric model in the field of parametric spectral analysis, modeling and forecasting.

The spectral properties of the PCRP are characterized by a two-frequency spectral density. For the considered class of processes, the spectral density is concentrated only along the straight lines $\omega_2 = \omega_1 - k\frac{2\pi}{T}$. Therefore, we can reduce the empirical spectral analysis of PCRP to the estimation of components $\hat{f}_k(\omega)$ or the variable spectral density $\hat{f}(\omega, t)$ whose Fourier expansion involves these components (Dragan et al. 1987).

We can form the statistics of the variable spectral density and spectral components using expressions (7), (8), and (10). Similarly to the spectral analysis of stationary random processes, we can derive consistent estimates by smoothing correlograms:

$$\hat{f}(\omega, t) = \frac{1}{2\pi} \int\limits_{-u_m}^{u_m} \hat{b}(t, u) k(u) e^{-i\omega u} du,$$

$$\hat{f}_k(\omega) = \frac{1}{2\pi} \int\limits_{-u_m}^{u_m} \hat{B}_k(u) k(u) e^{-i\omega u} du \tag{19}$$

The correlation window $k(u)$ is an even function $k(-u) = k(u)$ and $k(0) = 1$. For $|u| \geq u_m$, where u_m is the point of correlogram truncation $k(u_m) = 0$. For a given realization length θ, the variances of the estimates (19) decrease with the narrowing of the correlation window. Fluctuation components of estimate biases should display a similar behavior. However, by increasing u_m we can reduce the bias components that determine the resolving power of spectral analysis. Such contradictory tendencies in variations of estimate characteristics impede the choice of parameters θ and u_m. For PCRP with known or preset characteristics, we can ensure a substantiated choice of these parameters from evaluated characteristics of the statistical accuracy of estimation. The employed approach allows us to provide recommendations for processing PCRP realizations of specific types. This approach is favorable for revealing general properties of spectral estimates. Note that the results of comprehensive investigations of the properties of spectral estimates for stationary random processes provide the basis for the empirical spectral analysis of PCRP at the initial stage of studies.

The methods of coherent and component estimation are also suitable for the statistical analysis of PCRP-related processes, such as vector-PCRP, bi-PCRP. and poly-PCRP (Javors'kyj 1986, 1987). However, generally, characteristics of bi- and poly-PCRP can be estimated only with using of the component method. Coherent averaging is applicable only when data sampling is performed with a small time interval that contains an integer number of periods of other rhythms. If this condition is not satisfied, then we can employ coherent data sampling, where the realization length cannot be matched with time scales of variations in probabilistic characteristics, only to estimate additive components. If we can separate time intervals that include a sufficient number of smaller periods and temporal variations of characteristics corresponding to the larger period are insignificant within these intervals of time, then coherent averaging within such time intervals allow us to evaluate sliding estimates for the characteristics of bi-and poly-PCRP with a satisfactory accuracy. Such an estimation procedure assumes that these generalizations of PCRP can be represented as PCRP with slowly varying characteristics.

Both coherent and component estimates of the characteristics of polyrhythmic processes are biased, and their biases are determined, to a great extent, by the difference of the relevant correlation periods. If this difference is small, then, to ensure the required resolving power, we should choose the processed realization fragment in such a way that its length should be much greater than that required to ensure the smallness of biases when each rhythm is processed separately. It is not always the case that we can meet this requirement. Then, we encounter an urgent problem of using processing methods with a higher selectivity (Javors'kyj 1988; Mikhailyshyn and Javors'kyj 1990; Zabolotnyj et al. 2000; Yavorskyj et al. 2011), in particular, the least-squares method.

Above, we considered the properties of continuous estimates of probabilistic characteristics. In the overwhelming majority of problems of practical importance, we deal with time series that consist of discrete sequences of values $\xi\,(nh)$ where h is the sampling interval. We should analyze the influence of the sampling interval on the estimation quality both from the viewpoint of revealing reliable changes in the estimates for the mean, correlation function, and the spectral density in time domain and in the context of the possibility to make reliable conclusions concerning the dependences of the correlation function and correlation components on the lag and the sensitivity of spectral characteristics to the frequency. To obtain intermediate time-domain estimates with using the coherent method, it would be appropriate to employ a trigonometric interpolation (Javors'kyj 1987). Interpolated estimates are unbiased if the sampling step h satisfies the condition $h \leq T / (2M + 1)$, where are the numbers of higher order components of characteristics being evaluated. Simultaneously, this condition ensures the unbiasedness sample estimates for the components (Javors'kyj 1985). If this condition is not met component estimates are perturbed by the effect of overlapping. The variances of estimates depend on the correlation properties of PCRP. The sampling interval should be chosen in such a manner as to ensure the smallness of the differences between the reliability of discrete estimates and the reliability of the corresponding continuous estimates.

If the correlation period is not multiple of the sampling interval h, then we should apply the component method of estimation, because coherent data sampling in such a situation would result in the accumulation of errors. The use of the component method simultaneously solves the interpolation problem. For $h \leq T / (2M + 1)$, the relation between the qualities of coherent and component discrete estimates is close to that characteristic of continuous processing (Isayev and Javors'kyj 1995).

If the sampling interval is not matched properly with the temporal structure of a PCRP, then the sample estimates of spectral components are also distorted. In this case, the estimates of the components $f_l\,(\omega)$ take the values $f_{l+qL}\,(\omega)$, $q \in Z$, $L = T/h$. The estimation of the spectral components can be also accompanied by additional errors due to the overlapping of the values of the same component at the frequencies $\omega \pm 2\pi k / \Delta u$, where Δu is the lag discretization interval. To reduce these errors, we should choose the corresponding step Δu. These errors are insignificant if Δu is reduced to such an extent that the values of the spectral components at the frequencies $\omega \pm 2\pi k / \Delta u$ are sufficiently small. The step Δu also influences the variance of the estimates. If the values of the spectral characteristics are large outside the interval $[-\pi / \Delta u, \pi / \Delta u]$ than the statistical accuracy of discrete estimates is considerably lower continuous estimates.

To eliminate overlapping errors arising when we process time series using the bi-PCRP methods, we should match the sampling step with the numbers of highest order components for both correlation periods. Naturally, evaluating the estimates for the mean and the correlation function, we should satisfy different requirements to the samplings step. If this step is such that $h \leq T_1 / (2N_1 + 1)$ and $h \leq T_2 / (2N_2 + 1)$

where N_i are the numbers of highest order components for each of the periods, then the errors that arise when continuous averaging is replaced by discrete averaging are determined by the difference between the relevant integral transforms of the correlation components and the corresponding integral sums for the given sample step.

In many situations that occur in reality, time intervals that separate different values of series of empirical data are not only variable but also random. This circumstance necessitates the construction and investigation of PCRP statistics in the case of stochastic discretization. Analysis of such series can be based on the modified coherent and component methods (Javors'kyj 1985; Kostjukov et al. 1987). If the values of a time series are obtained in the so-called stationary discretization regime, then both of these methods provide estimates whose quality is only slightly lower than the quality of estimates evaluated from equidistant time series. In such a situation, the component method displays a higher sensitivity to the deviations of the properties of discretization flows from the stationary flow.

4 Detection of Hidden Periodicities

To apply the methods of statistical analysis based on PCRP and their generalizations, we should preliminarily determine the correlation periods for the processes under investigation. In many cases, we can make a decision concerning the values of these periods based on the analysis of the physical nature of the phenomenon under study. Specifically, when studying daily and annual variations of geophysical processes, we can find an obvious solution to the problem associated with the correlation period. Astronomical factors determine a poly-rhythmic character of tidal oscillations of the sea level, sea currents, and internal waves. In this case, the correlation periods are determined from the well-known poly-harmonic representation of the potential of tide-generating forces. When we investigate rhythmic variations of stochastic intrinsic oscillations that occur in various dynamical systems, the problem of determining the correlation period ceases to be trivial. In such situations, to solve this problem, we should develop methods of estimation that would be adequate to the accepted models (Javors'kyj 1984, 1985; Javors'kyj and Mikhailyshyn 1996).

The problem of the detection of hidden periodicities, which is a classical problem of mathematical statistics of random processes, was formulated back in the nineteenth century (Serebrjennikov and Pervozvansk 1965; Yaglom 1981). Initially, the solution of this problem was reduced to the evaluation of parameters of a periodic or nearly periodic deterministic function. Subsequently, the detection of hidden periodicities was transformed into the problem of searching for reliable peaks of the power spectral density for stationary random processes.

In terms of the PCRP model, hidden periodicities do not necessarily manifest themselves as peak values of the spectral density (Dragan et al. 1987; Yaglom 1981). Within the framework of the PCRP model, the detection of periodicities is formulated as a problem of determining the period of temporal variations in the estimates of probabilistic characteristics of stochastic oscillations, such as the mean, correlation function and variable spectral density. Such an approach separates the search for the period of regular oscillations from the determination of the recurrence period of correlations.

Based on the representation (4), we can reduce the problem of the determination of the PCRP correlation period to the problem of parametric estimation. In the case of a Gaussian PCRP, we can employ the maximum likelihood method. The corresponding estimate can be represented in the form of a power series in a certain small parameter ε.

$$\hat{T} = T_0 + \varepsilon T_1 + \varepsilon^2 T_2 + \cdots \qquad (20)$$

In the first-order approximation for ε, this estimate is unbiased, and the variance of this estimate coincides with the variance of the efficient estimate. If we increase the realization length being processed, the parameter ε decreases simultaneously with the accuracy of the first-order approximation. Therefore, the maximum-likelihood estimate of the period is asymptotically unbiased and asymptotically efficient. The efficiency of this estimate is due to the use of apriori data concerning the probabilistic structure of the PCRP. At the initial stage of investigation, such data are usually lacking, which makes us employ less efficient methods that do not require any apriori information. Development of such methods can be based on the concepts of the above-considered methods of coherent and component estimation. Both coherent and component statistics are characterized by certain resonant properties with respect to the correlation period. These statistics reach their extreme at points T that asymptotically tend to the true values of the period. Specifically, the functionals of the mean can be written as:

$$\hat{m}(t) = \frac{1}{2N+1} \sum_{n=-N}^{N} \xi(t+n\tau), \qquad (21)$$

$$\begin{Bmatrix} \hat{m}_k^c \\ \hat{m}_k^s \end{Bmatrix} = \frac{1}{\theta} \int_0^{\theta} \xi(t) \begin{Bmatrix} \cos \\ \sin \end{Bmatrix} \Lambda_k t \, dt, \quad \Lambda_k = k2\pi/\tau, \qquad (22)$$

$$m(t,\tau) = \sum_{n=-N_1}^{N_1} \hat{m}_k(\tau) e^{i\Lambda_k t}. \qquad (23)$$

Analogously to relations (21)–(23), we can also define the functionals of the correlation function (Drabych et al. 2000). Similarly to the maximum likelihood method, the estimates of the period are determined from the nonlinear equation:

$$\frac{dS(\tau)}{d\tau} + \varepsilon \frac{dN(\tau)}{d\tau} = 0,$$

where $S(\tau)$ and $N(\tau)$ are the regular and fluctuation components of the functionals (21)–(23) respectively and ε is a small parameter. Solutions to this equation can be represented in the form (20). If the correlation function decreases with the growth in the lag, then the small parameter, which is defined by the expression $\varepsilon = \sqrt{\left[EN^2\left(T\right)\right]/S\left(T\right)}$, tends to zero as $\theta \to \infty$. Therefore, the estimates of the period are asymptotically unbiased and consistent.

The advantage of the statistics (21)–(23), similarly to the statistics of correlation characteristics defined by analogy with (21)–(23), is associated with the fact that, in addition to the correlation period, these statistics allow us to simultaneously determine the characteristics of PCRP.

The developed procedure for the detection and analysis of hidden periodicities was employed in the investigation of wind waves, swell, Wolf number series, geomagnetic pulsations Pel and Pc3, annual and interyearly rhythmic variations of hydrometeorological processes and vibroacoustic signals. Based on this procedure, we revealed qualitatively new features of the probabilistic structure of rhythmic variations of the above-specified processes (Dragan et al. 1987; Mikhailyshyn et al. 1990; Dragan and Javors'kyj 1982; Yavorskyj et al. 2012; Gudzenko 1961).

The results of our studies clearly demonstrate that the methods for the detection of hidden periodicities based on functionals (21)–(23) allow us also to make conclusions concerning the applicability boundaries of models of rhythmic variations in the form of PCRP and their generalizations. The proposed approach naturally extends the problem of hidden periodicities searching on the basis of a more logical and comprehensive notion of rhythmic of stochastic oscillations, which, apart from other advantages, removes a severe limitation associated with the requirement that the considered oscillation process should be stationary.

5 Conclusions

Relying on the results of the performed investigations, we developed, in collaboration with our colleagues, the fundamentals of the theory of the statistical analysis of rhythmic signals on the basis of models in the PCRP form and their generalizations. Within the framework of the spectral-correlation theory of nonstationary random processes of these classes, we substantiated a general approach to the investigation of the stochastic recurrence in phenomena of different nature. It is shown that, using the first-and second-order characteristics of such processes, we can reveal substantial and closely interrelated features of rhythmic variations, such as the periodicity and randomness. Eventually, these features reflect stochastic amplitude and phase modulation of signals. We developed and investigated the methods for the estimation of probabilistic characteristics of PCRP and their generalizations, including the coherent and component methods, the least-squares method and methods of linear filtration. We developed a general approach to the problems of estimation and construction of optimal estimates for PCRP characteristics. On the basis of the PCRP model, we developed methods for the detection and analysis of hidden periodicities. We also created software for the statistical processing of signals with the stochastic recurrence. With the use of these means, we revealed previously unknown properties of rhythmic variations of several physical processes that occur on the Earth, in the atmosphere, in the ocean, in the ionosphere and in various technological systems. The proposed parametric models of stochastic oscillations provide the background for the performing and designing of experimental statistical investigations, simulation and forecasting of various processes and recognition and diagnostics of the states of dynamic systems that generate these processes.

References

Antoni J (2009) Cyclostationarity by examples. Mech Syst Signal Process 23:987–1036

Aschoff Y (ed) (1981) The biological rhythms. Plenum Press, New York

Bennet WR (1958) Statistics of regenerative digital transmission. Bell System Techn Jorn 37: 1501–1542

Bjelyshev A, Klevantsov Y, Rozhkov V (1983) Probabilistic analysis of sea currents. Gidrometeoizdat, Leningrad (in Russian)

Drabych O, Mykhajlyshyn V, Javors'kyj I (2000) Determination of correlation period of the periodically correlated random processes using covariation transformations. Vidbir i Obrobka Informatsiji 14(90):47–52 (in Ukrainian)

Dragan Y (1980) The structure and representation of the stochastic signal models. Naukova Dumka, Kiyev (in Russian)

Dragan Y (1972) On biperiodically correlated random processes. Otbor i Peredacha Informatsiyi 34:12–14 (in Russian)

Dragan Y, Javors'kyj I (1975) About representation of sea waves by periodically correlated random processes and their statistical processing methods. Methods of representation and apparatus analysis of random processes and fields. XIII Ail-Union Symposium, Leningrad, pp 29–33 (in Russian)

Dragan Y, Javors'kyj I (1982) Rhythmic of Sea Waving and Underwater Acoustic Signals. Naukova Dumka, Kyiv (in Russian)

Dragan Y, Rozhkov V, Javors'kyj (1987) The methods of probabilistic analysis of oceanological rhythmic. Gidrometeoizdat, Leningrad (in Russian)

Franks LE (1969) Signal theory. Prentice-Hall, Englewood Cliffs

Gardner WA (1994) Cyclostationarity in communications and signal processing. In: Gardner W (ed) IEEE, New York

Gardner WA, Franks LE (1975) Characterization of cyclostationary random signal processes. IEEE Trans Inf Theory IT-21:4–14

Gardner WA (1985) Introduction to random processes with applications to signals and systems. Macmillan, New-York

Gardner WA (1986) The spectral correlation theory of cyclostationary time-series. Signal Process 1:3–36

Gladyshev E (1959) Periodically and almost periodically correlated processes with continuous time. Teoriya Veroyatnostei i yeyo Primenenie 3(2):84–189 (in Russian)

Groisman P (1977) The estimate of autocorrelation matrixes of the precipitation time series considered as periodically correlated random processes. Trudy GGL 247:119–127 (in Russian)

Gruza G (1982) Some general problems of time series statistical analysis in climatology. In: Proceedings of all-union scientific research institute of hydrometeorologi-cal information—world data centre 83, Obninsk, pp 3–9 (in Russian)

Gudzenko L (1961) The generalization of the ergodic theorem for nonstationarv random processes. Izvestia Vysshikh Uchebnykh Zavedenij Ser Radiofizika 4(2):265–274 (in Russian)

Gudzenko L (1959) On periodically nonstationary processes. Radiotekhnika i Elektronika 6(6):1020–1040 (in Russian)

Gudzenko L (1959) The small fluctuation in essentially nonlinear auto-oscillation system. Dokl Akad Nauk USSR 125(1):62–65 (in Russian)

Hurd HL (1969) An investigation of periodically correlated stochastic processes. Ph.D. dissertation. Duke University Department of Electrical Engineering

Hurd HL (1989) Nonparametric time series analysis for periodically correlated processes. IEEE Trans Inf Theory 35:350–359

Isayev I, Javors'kyj I (1995) Component analysis of the time series with rhythmical structure. Izvestia Vysshykh Uchebnykh Zavedeniy Ser Radioelektronika 38(1):34–45 (in Russian)

Javors'kyj I, Mikhailyshyn V (1996) Probabilistic models and investigation of hidden periodicities. Appl Math Lett 9(2):21–23

Javors'kyj I (1984) The application of Buys-Ballot scheme for statistical analysis of rhythmical signals. Izvestiya Vysshykh Uchebnykh Zavedenij, ser. Radio-elektronica 27(11):31–37 (in Russian)

Javors'kyj I (1985) The computation of characteristics of periodically correlated random processes by data random selection. Thesis of reports of I acoustical seminar on modelling, algorithms and reception of decision. Moscow, pp 44–40 (in Russian)

Javors'kyj I (1987) Biperiodically correlated random processes as model of bi-rhythmic signals. In: All-Union Conference on Information Acoustics, Moscow, pp 6–10 (in Russian)

Javors'kyj I. (1988) The least squares method for statistical analysis of signals with rhythmical structure. In: II all-union acoustical seminar on modelling, algorithms and reception of decisions, Moscow, pp 5–6 (in Russian)

Javors'kyj I (1985) The component sample estimates of probabilistic characteristics of periodically correlated random processes. Otbor i Peredacha Informatsivi 72:17–27 (in Russian)

Javors'kyj I (1985) On statistical analysis of periodically correlated random processes. Radiotekhnika i Electronika 6:1096–1104 (in Russian)

Javors'kyj I (1986) Statistical analysis of biperiodically correlated random processes. Otbor i Peredacha Informatsiyi 73:12–21 (in Russian)

Javors'kyj I (1987) The statistical analysts of vector periodical correlated random processes. Otbor i Peredacha Informatsiyi 76:3–12 (in Russian)

Javors'kyj I (1987) The interpolation of the estimates of periodicallly correlated random processes. Avtomatika I:36–41 (in Russian)

Javorskyj I, Kravets I, Isayev I (2006) Parametric modeling of periodically correlated random processes by their representation through stationary random processes. Radio Electron Commun Syst 49(11):23–29

Javorskyj I, Isaev I, Zakrzewski Z, Brooks SP (2007) Coherent covariance analysis of periodically correlated random processes. Signal Process 8(1):13–32

Javorskyj I, Isayev I, Majewski J, Yuzefovych R (2010) Component covariance analysis for periodically correlated random processes. Signal Process 90:1083–1102

Javorskyj I, Leskow J, Kravets I, Isayev I, Gajecka E (2012) Linear filtration methods for statistical analysis of periodically correlated random processes—Part I: coherent and component methods. Signal Process 92:15591566

Javorskyj I, Isayev I, Kravets I, (2007) Algorithms for separating the periodically correlated random processes into harmonic series representation. In: Proceedings of 15th European signal processing conference (EUSIPCO 2007), Poznan, pp 1857–1861

Javorskyj I, Yuzefovych R, Krawets I, Zakrzewski Z (2011) Least squares method in the statistic analysis of periodically correlated random processes. Radoi Electron Commun Syst 54 (1):45 59

Jaworskyj IM, Drabych PP, Kravets IB, Matsko II, (2011a) Metod of vibration diagnostics of initial stages of rotation systems damage. Mater Sci 47(2):264–271

Jaworskyj I, Leskow J, Kravets I, Isayev I, Gajecka E (2011b) Linear filtration methods for statistical analysis of periodically correlated random processes—Part II: harmonic series representation. Signal Process 91(2506):2519

Kiselyeva T, Chudnovsky A (1968) Statistical investigation of the diurnal cycle of the air temperature. Bull Nauchno-tekhnicheskoi Informatsiyi po Agronom Fizike 11:17–38 (in Russian)

Kolyesnikova V, Monin A (1965) On spectrum of meteorological field oscillations. Izvestiya AN USSR, Ser. Fizika Atmosfery i Okeana 1(7):653–669 (in Russian)

Koronkevich O (1957) The linear dynamic systems under action of the random forces. Naukovi Zapyskj Lvivskoho Universytetu 44(8):175–183 (in Ukrainian)

Kostjukov Y, Uljanich I, Mezentsev V, Javorskyj I (1987) The analysis of nonequidistant time series of Gulf of Riga temperature and salinity. In: Ail-Union Scientific Research Institute of Hydro meteorological World Data Centre, vol 134, pp 87–97 (in Russian)

Kozel S (1959) The transformation of the periodically nonstationary fluctuations by linear filter. Trudy Moskovskoho Fiziko-tekhnicheskoho Instituta 4:10–l6 (in Russian)

Kravets IB (2012) Parametric models of cyclostationary signals. Radio Electron Commun Syst 55(6):257–267

Malakhov A (1968) The fluctuation in autooscillation systems. Nauka, Moscow (in Russian)

Mamontov N (1968a) The standard deviation and coefficients of skewness of air temperature in south-east of West-Siberia plain. Trudy NIIAK 54:29–34 (in Russian)

Mamontov N (1968b) The investigation of distribution statistics of the relative air humidity. Trudy NIIAK 54:18–28 (in Russian)

Mezentsev V, Javors'kyj I (1988) The properties of the optimum estimates of rhythmic signal probabilistic characteristics. Thesis of the reports of All-union workshop On signal processing. Uljanovsk Polytechnical Institute, Uljanovsk, pp 16–18 (in Russian)

Mikhailyshyn V, Fligel S, Javors'kyj I (1990) Statistical analysis of wave packets of geomagnetic pulsations Pcl type by the method of periodically correlated random processes. Geomagnetizm i Aeronomia 30(5):757–764 (in Russian)

Mikhailyshyn V, Fligel S, Javors'kyj (1990) The probabilistic model of the signal periodicity of geomagnetic pulsations. Pel. The investigation of the structure and wave properties near-earth plasma. Nauka, Moscow, pp 76–88 (in Russian)

Mikhailyshyn V, Javors'kyj I (1990) LSM-analysis while identifying poly-rhythmic structure of stochastic signals. In: Proceedings of the first international conference on information technologies for image analysis and pattern recognition, Lviv, pp321–325

Mishchenko Z (1966) The air temperature diurnal cycle and its agroclimatic significance. Gidrometeoizdat, Leningrad (in Russian)

Mishchenko Z (1960) The air temperature diurnal cycle and plant termo-periodicity. Trudy GGO 91:15–28 (in Russian)

Mykhailyshyn VYu, Yavors'kyi IM, Vasylyna YuT, Drabych OP (1997) Probabilistic models and statistical methods for the analysis of vibrational signals in the problems of diagnostics of machine. Mater Sci 33(5):655–672

Myslovich M, Pryimak N, Shcherbak L (1980) Periodically correlated random processes in problems of acoustic information processing. Znaniye, Kiyev (in Russian)

Ogura H (1971) Spectral representation of periodic nonstationary random processes. IEEE Trans Inf Theory 17(2):143–149

Papoulis A (1983) Random modulation: a review. IEEE Trans Acoust Speech Signal Process 31(1):96–105

Poljak I (1978) Methods of analysis of random processes and fields in climatology. Gidrometeoizdat, Leningrad (in Russian)

Romanenko A, Sergeyev G (1968) The problems of the applied analysis of random processes. Sov Radio, Moscow (in Russian)

Rytov S (1976) An introduction to statistical radio-physics. P.1. Nauka, Moscow (in Russian)

Serebrjennikov M, Pervozvansk A (1965) The hidden periodicities detecting. Nauka, Moscow (in Russian)

Stratonovich R (1961) Selected problems of the fluctuation theory in radioengneering. Sov Radio, Moscow (in Russian)

Tikhonov V (1956) When nonstationary random process can be substituted for stationary process. Zhurnal Teoreticheskoi i Ekspcrimentalnoi Fiziki 31(9):2057–2059 (in Russian)

Voichishyn K, Dragan Y (1971) On simple stochastic model of the natural rhythmic processes. Otbor i Peredacha Informatsiyi 29:7–15 (in Russian)

Voichyshyn K (1975) The problems of statistical analysis of nonstationary (rhythmic) phenomena conformable to some geophysical object. Cand. Phys. and Math. Sci. Dissertation, Institute of Earth Physics, Moscow (in Russian)

Ya Dragan (1978) The harmonizability and the decomposition of the random processes with finite average power. Dokl Akademiyi Nauk USSR ser A 8:679–684 (in Russian)

Ya Dragan (1969) On periodically correlated random processes and systems with periodic parameters. Otbor i Peredacha Informatsiyi 22:27–33 (in Russian)

Ya Dragan (1970) On spectral properties of periodically correlated random processes. Otbor i Peredacha Informatsiyi 30:16–24 (in Russian)

Ya Dragan (1972) About basing of rhythmic stochastic model. Otbor i Peredacha Informatsiyi 31:12–21 (in Russian)

Ya Dragan (1972) The properties of the periodically correlated random process samples. Otbor i Peredacha Informatsiyi 33:9–12 (in Russian)

Ya Dragan (1975) On representation of periodically correlated random processes by stationary components. Otbor i Peredacha Informatsiyi 45:7–20 (in Russian)

Ya Dragan (1985) Periodic and periodically nonstationary random processes. Otbor i Peredacha Informatsiyi 72:3–17 (in Russian)

Yaglom A (1981) Correlation theory of stationary random functions. Gidrometeoizdat, Leningrad (in Russian)

Yavors'kyi I, Isaev I, Kravets I, Drabych P, Mats'ko I (2009) Methods for enhancement of the efficiency of statistical analysis of vibration signals from the bearing supports of turbines at thermal-electric power plants. Mater Sci 45(3):378–391

Yavorskyj IN, Kravets IB, Mats'ko IY (2011) Spectral analysis of stationary components of periodically correlated random processes. Radio Electron Commun Syst 54 8):451 463

Yavorskyj IN, Yuzefovych R, Kravets IB, Matsko IY (2012) Properties of characteristics estimators of periodically correlated random processes in preliminary determination of the period of correlation. Radio Electron Commun Syst 55(8):335–348

Zabolotnyj O, Mykhajlyshyn V, Javors'kyj I (2000) The least squares method of polyrhythmic statistical analysis. Dopovidi NAN Ukrainy 8:93–100 (in Ukrainian)

Zhukovsky E (1969) The investigation of the statistical characteristics of relative air humidity. Sbornik Trudov po Agronom Fizike 20:3–28 (in Russian)

Simulation Comparison of CBB and GSBB in Overall Mean Estimation Problem for PC Time Series

Anna E. Dudek and Paweł Potorski

Abstract In the chapter the performance comparison in the simulation study of the block bootstrap methods that can be used in the problem of the overall mean estimation of a PC time series is presented. Two block bootstrap techniques are considered: the Circular Block Bootstrap and the circular version of the Generalized Seasonal Block Bootstrap. The actual coverage probabilities of the bootstrap equal-tailed confidence intervals are calculated for a wide range of the block length choices and a few sample sizes. Moreover, the optimal values of the block lengths are pointed. In the most of the considered cases performance of CBB and GSBB is very comparable.

1 Introduction

Periodically correlated (PC) time series are frequently used to model data with periodic structure in many different fields like climatology (Bloomfield et al. 1994, 1995), economy (Dudek et al. 2013) and mechanical signals (Antoni 2009). Time series $\{X(t), t \in \mathscr{Z}\}$ is called PC if it has periodic mean and covariance functions

$$E\left(X(t+d)\right) = E\left(X(t)\right) \quad \text{and} \quad Cov\left(X(t+d), X(s+d)\right) = Cov\left(X(t), X(s)\right),$$

where d is the known period length. For more details we refer the reader to the book of Hurd and Miamee (2007).

The important subgroup of PC processes are PARMA (p, q) (periodic auto-regressive-moving-average) processes, which are of the form

A. E. Dudek (✉) · P. Potorski
AGH University of Science and Technology, al. Mickiewicza 30, 30-059 Krakow, Poland
e-mail: aedudek@agh.edu.pl

P. Potorski
e-mail: potorski@agh.edu.pl

F. Chaari et al. (eds.), *Cyclostationarity: Theory and Methods*,
Lecture Notes in Mechanical Engineering, DOI: 10.1007/978-3-319-04187-2_7,

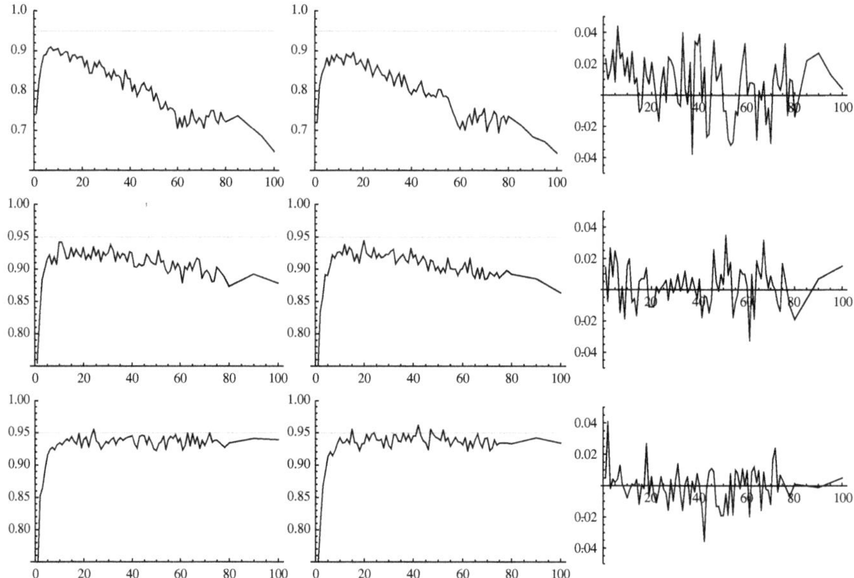

Fig. 1 First and second column: ACPs (*black lines*) for **M1** for $b \in \{1, 2, \ldots, 79, 80, 90, 100\}$ together with nominal coverage probability (*grey line*) for CBB and GSBB, respectively. Third column: differences between ACPs for CBB and GSBB. Rows 1–3 correspond to the sample size $n = 120$, $n = 480$ and $n = 1920$, respectively

$$X(t) = \sum_{j=1}^{p} \phi_j(t) X(t - j) + \sum_{k=1}^{q} \theta_k(t) \xi(t - k) + \sigma(t) \xi(t), \tag{1}$$

where $\phi_j(t) = \phi_j(t + T)$, $\theta_k(t) = \theta_k(t + d)$, $\sigma(t) = \sigma(t + d)$ for all $j = 1, \ldots, p$, $k = 1, \ldots, q$ are periodic coefficients, and $\xi(t)$ is mean zero white noise with variance equal to one.

In the sequel we use an alternative parametrization introduced by Jones and Brelsford (1967) to reduce the number of parameters required to represent PARMA model.

$$\phi_j(t) = a_{j,1} + \sum_{m=1}^{\lfloor d/2 \rfloor} a_{j,2m} \cos(2\pi m t/d) + \sum_{m=1}^{\lfloor d/2-1 \rfloor} a_{j,2m+1} \sin(2\pi m t/d), \quad j = 1, \ldots, p,$$

$$\theta_k(t) = b_{k,1} + \sum_{m=1}^{\lfloor d/2 \rfloor} b_{k,2m} \cos(2\pi m t/d) + \sum_{m=1}^{\lfloor d/2-1 \rfloor} b_{k,2m+1} \sin(2\pi m t/d), \quad k = 0, \ldots, q,$$

where $\theta_0(t) = \sigma(t)$.

The reduction of the number of parameters can be obtained by restricting the number of frequencies in the Fourier series. Then, the estimates of the Fourier coefficients

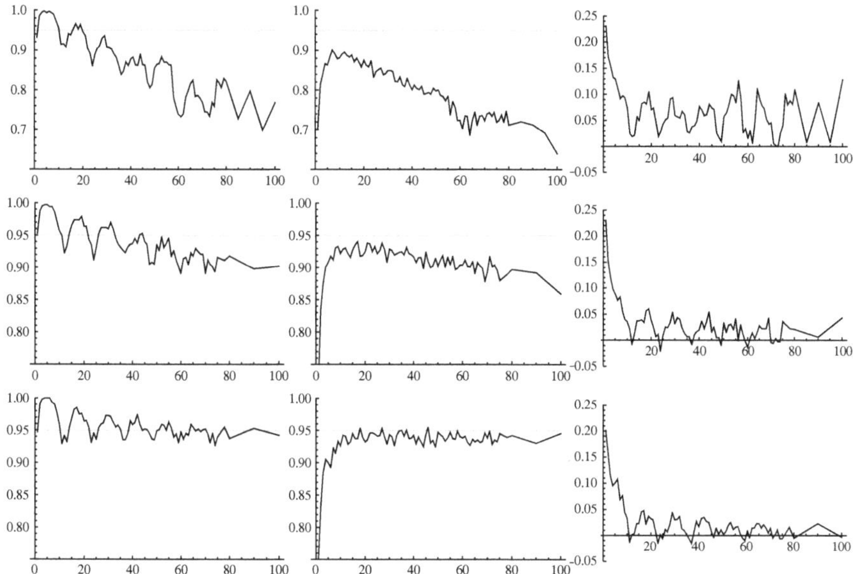

Fig. 2 First and second column: ACPs (*black lines*) for **M2** for $b \in \{1, 2, \ldots, 79, 80, 90, 100\}$ together with nominal coverage probability (*grey line*) fcr CBB and GSBB, respectively. Third column: differences between ACPs for CBB and GSBB. Rows 1–3 correspond to the sample size $n = 120$, $n = 480$ and $n = 1920$, respectively

are calculated using the maximum likelihood method. Since the above transformation is one-to-one the solution can be transformed to $\{\phi_j(t), \theta_k(t)\}$.

The overall mean estimation problem for PC time series is quite well described in the literature. The resampling techniques are often used to construct the confidence interval for the parameter. So far the consistency of a few block bootstrap methods was proved. The first result was obtained by Synowiecki (2007) for the Moving Block Bootstrap Method (MBB). Although, MBB is a very well known and widely applicable it does not respect the periodic structure of the considered data. As a result it cannot be used for example in the problem of the seasonal means estimation. To address this issue the block bootstrap methods that preserve the periodicity of the original data were proposed. In Politis (2001) and Chan et al. (2004) the Seasonal Block Bootstrap (SBB) and in the Periodic Block Bootstrap (PBB) were introduced, respectively. The consistency of SBB for the overall mean was shown in Synowiecki (2008). Unfortunately, PBB turned out to be inconsistent unless the period length is growing together with the sample size (for more details see Leśkow and Synowiecki 2010). Finally, Dudek et al. (2014) proposed the Generalized Seasonal Block Bootstrap (GSBB), which is the generalization of PBB and SBB. In contrary to those PBB and SBB, GSBB allows for quite arbitrary choice of the block length. Moreover, the method is consistent for the overall mean and the seasonal means.

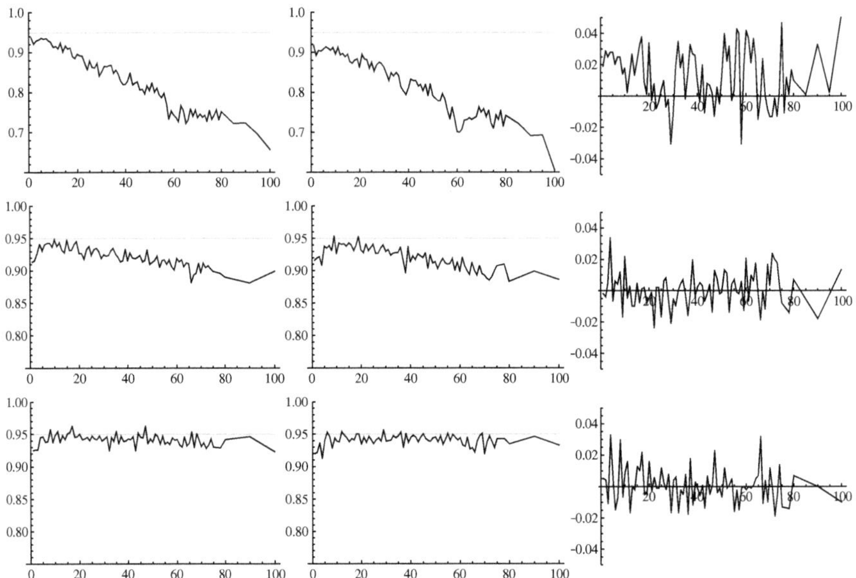

Fig. 3 First and second column: ACPs (*black lines*) for **M3** for $b \in \{1, 2, \ldots, 79, 80, 90, 100\}$ together with nominal coverage probability (*grey line*) for CBB and GSBB, respectively. Third column: differences between ACPs for CBB and GSBB. Rows 1–3 correspond to the sample size $n = 120$, $n = 480$ and $n = 1920$, respectively

The aim of this chapter is to compare the performance of MBB and GSBB used for construction of the pointwise confidence intervals for the overall mean of the PC time series. It is known that MBB has the maximal degree of overlap among blocks and GSBB has (almost) full overlap in its successive blocks (see Dudek et al. 2014). On the other hand, the practitioner would like to use one method to perform all necessarily calculations. In this case the advantage of GSBB is noticeable as for example it provides the simultaneous confidence intervals for the seasonal means (Dudek et al. 2014).

2 Block Bootstrap Methods

In this section we recall the idea of MBB and GSBB. We decide to restrict only to the circular versions of these methods to avoid the edge effects. The circular version of MBB called the Circular Block Bootstrap (CBB) was introduced in Politis and Romano (1992).

Let $X(1), \ldots, X(n)$ be a sample from PC time series. By B_i we denote the block of observations that has the length b and starts with $X(i)$ i.e. $B_i = (X(i), \ldots, X(i + b - 1))$.

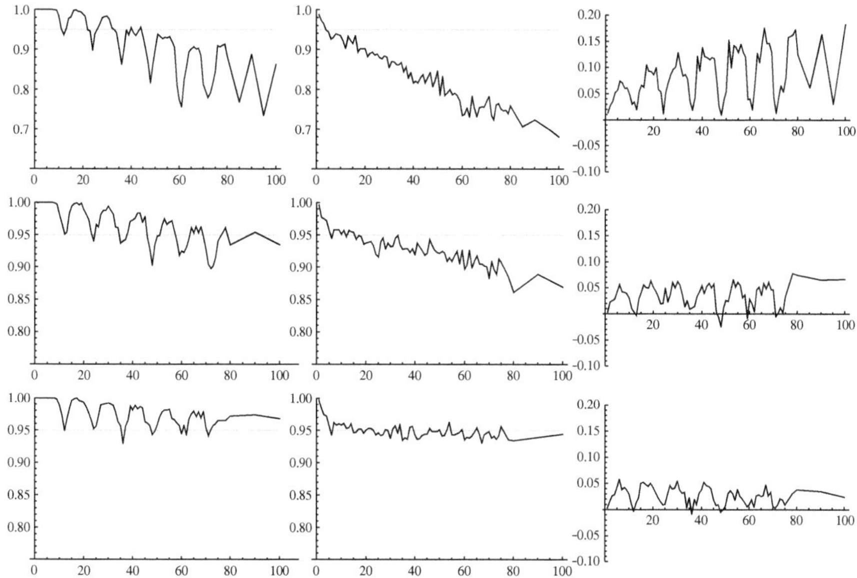

Fig. 4 First and second column: ACPs (*black lines*) for **M4** for $b \in \{1, 2, \ldots, 79, 80, 90, 100\}$ together with nominal coverage probability (*grey line*) for CBB and GSBB, respectively. Third column: differences between ACPs for CBB and GSBB. Rows 1–3 correspond to the sample size $n = 120, n = 480$ and $n = 1920$, respectively

If any time index i is greater than the sample size n we take the observation $i - n$ instead.

Moreover, without loss of generality and to simplify the notation we assume that the sample size n is an integer multiple of the block length b ($n = lb$) and is an integer multiple of the period length d ($n = wd$).

CBB Algorithm:

1. Choose a (positive) integer block size $b(< n)$.
2. For $t = 1, b + 1, 2b + 1, \ldots, (l - 1)b + 1$, let

$$\left(X^*(t), \ldots, X^*(t + b - 1)\right) = B_{k_t},$$

where k_t is iid from a discrete uniform distribution

$$P\left(k_t = s\right) = \frac{1}{n} \quad \text{for} \quad s = 1, \ldots, n.$$

This means that we are choosing randomly with the replacement l blocks of the length b. Each block is selected with probability $1/n$.
3. Join the l blocks B_{k_t} to get the bootstrap sample.

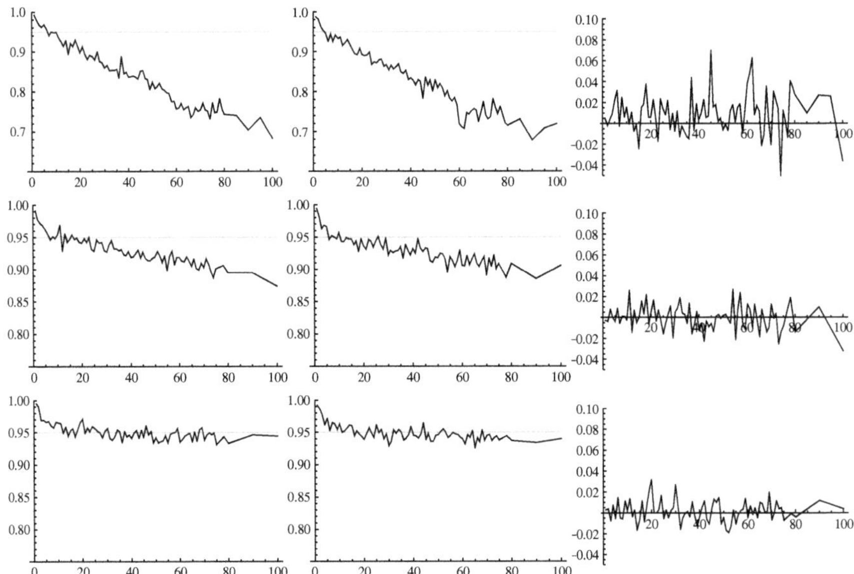

Fig. 5 First and second column: ACPs (*black lines*) for **M5** for $b \in \{1, 2, \ldots, 79, 80, 90, 100\}$ together with nominal coverage probability (*grey line*) for CBB and GSBB, respectively. Third column: differences between ACPs for CBB and GSBB. Rows 1–3 correspond to the sample size $n = 120$, $n = 480$ and $n = 1920$, respectively

Circular GSBB Algorithm:

1. Choose a (positive) integer block size $b(< n)$.
2. For $t = 1, b + 1, 2b + 1, \ldots, (l - 1)b + 1$, let

$$\left(X^*(t), \ldots, X^*(t + b - 1)\right) = B_{k_t},$$

where k_t is iid from a discrete uniform distribution

$$P\left(k_t = t + vd\right) = \frac{1}{w} \quad \text{for} \quad v = 0, 1, \ldots, w - 1.$$

3. Join the l blocks B_{k_t} to get the bootstrap sample.

If the condition $n = lb$ does not hold it is enough to select an additional block (independently of the chosen algorithm) and cut it to get the bootstrap sample of the same size as the original sample. Moreover, if $n \neq wd$ some of the blocks in GSBB will have higher probability of being chosen than others.

In the next section we present results of the performed simulations.

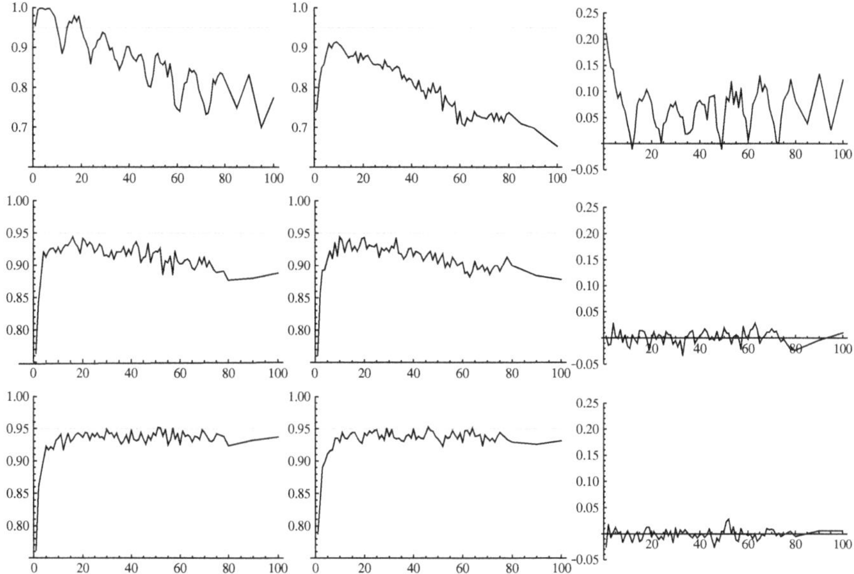

Fig. 6 First and second column: ACPs (*black lines*) for **M6** for $b \in \{1, 2, \ldots, 79, 80, 90, 100\}$ together with nominal coverage probability (*grey line*) for CBB and GSBB, respectively. Third column: differences between ACPs for CBB and GSBB. Rows 1–3 correspond to the sample size $n = 120, n = 480$ and $n = 1920$, respectively

3 Simulation Study

Our aim is to compare the performance of CBB and GSBB used to construct the bootstrap equal-tailed pointwise confidence intervals for the overall mean of PC time series. We consider a few examples of time series with periodic structure and calculate the actual converge probability (ACP) in each case. Moreover, for each example we take three values of the sample size n, namely 120, 480, 1920. The period length d is always equal to 12 and the number of bootstrap samples is $B = 500$. The nominal coverage probability is 95 % and to calculate the ACPs 1000 iterations are performed. The considered set of block length values is $\{1, 2, \ldots, 79, 80, 90, 100\}$. The considered PC time series are of the form:

M1: $X(t) = 0.5X(t-1) + \sin(2\pi t/d)\,\varepsilon(t),$

M2: $X(t) = \cos(2\pi t/d) + 0.5X(t-1) + \sin(2\pi t/d)\,\varepsilon(t),$

M3: $X(t) = 0.5\sin(2\pi t/d)\,X(t-1) + \sin(2\pi t/d)\,\varepsilon(t),$

M4: $X(t) = \cos(2\pi t/d) - 0.3\sin(2\pi t/d)\,\varepsilon(t-1) + \sin(2\pi t/d)\,\varepsilon(t),$

M5: $X(t) = -0.3\sin(2\pi t/d)\,\varepsilon(t-1) + \sin(2\pi t/d)\,\varepsilon(t),$

M6: $X(t) = 10\sin(2\pi t/d) + 0.45X(t-1) + 10\sin(2\pi t/d)\,\varepsilon(t),$

M7: $X(t) = -2\sin(2\pi t/d) + 10\sin(2\pi t/d)\,Y(t),$

M8: PARMA(2,1) with the nonzero coefficients: $a_{1,1} = 0.8$, $a_{1,2} = 0.3$, $b_{0,1} = -0.7$ and $b_{0,2} = -0.6$,

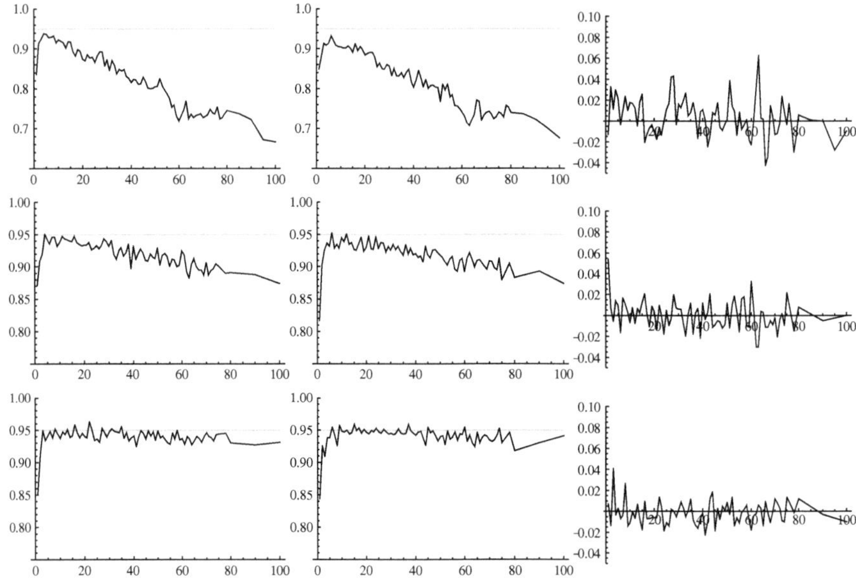

Fig. 7 First and second column: ACPs (*black lines*) for **M7** for $b \in \{1, 2, \ldots, 79, 80, 90, 100\}$ together with nominal coverage probability (*grey line*) for CBB and GSBB, respectively. Third column: differences between ACPs for CBB and GSBB. Rows 1–3 correspond to the sample size $n = 120$, $n = 480$ and $n = 1920$, respectively

M9: PARMA(2,1) with the nonzero coefficients: $a_{1,1} = 0.4$, $a_{1,2} = 0.6$, $b_{0,1} = -0.5$ and $b_{0,2} = -0.2$,

M10: PARMA(1,1) with the nonzero coefficients: $a_{1,1} = 0.4$, $a_{1,2} = 0.6$ and $b_{0,1} = -0.5$,

where $\{\varepsilon(t)\}_{t \geq 1}$ is a sequence of independent standard normal distribution random variables. The initial observations in each model are generated as standard normal random variables. Moreover, $Y(t)$ in **M7** is a zero-mean stationary sequence of the form $Y(t) = 0.45Y(t-1) + \varepsilon(t)$. Three PARMA time series (**M8–M10**) were obtained using 'makeparma' procedure provided by the R package 'perARMA' (see Dudek et al. 2013). We chose **M1–M10** for our consideration because these types of PC models can be met in many different applications. For example PARMA time series were used for climatology data in Bloomfield et al. (1994) and (1995).

The results are presented in Figs. 1, 2, 3, 4, 5, 6, 7, 8, 9, 10. For each case we provide ACP curve and additionally we calculate the differences between the ACPs obtained with CBB and GSBB. In Table 1 we present the optimal block length choices together with the ACP values obtained in these cases. If the same performance was observed for a few different block lengths we list them all.

One may note that for **M1** and **M8** values of ACP are too low for $n = 120$ and $n = 480$. For **M8** independently of the block bootstrap method the highest value of ACP is about 10% and 5% too low for $n = 120$ and $n = 480$, respectively. For all other cases

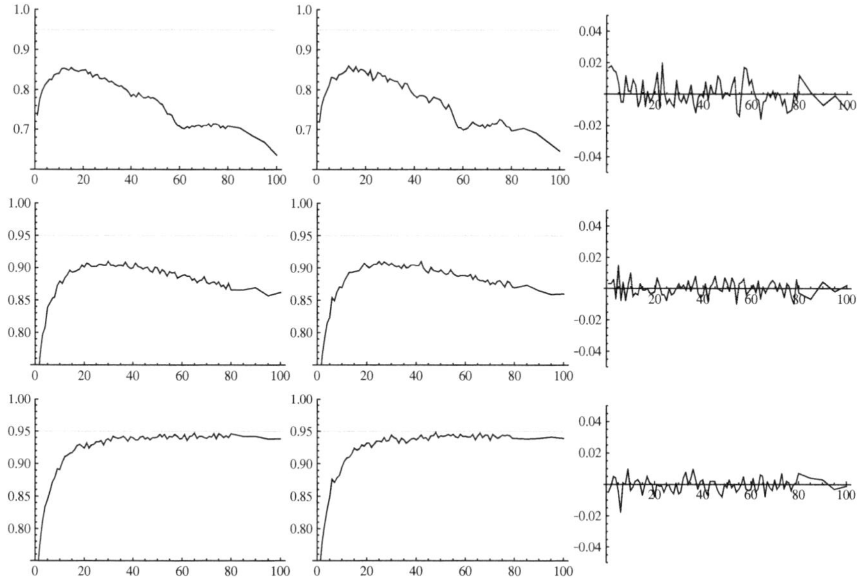

Fig. 8 First and second column: ACPs (*black lines*) for **M8** for $b \in \{1, 2, \ldots, 79, 80, 90, 100\}$ together with nominal coverage probability (*grey line*) for CBB and GSBB, respectively. Third column: differences between ACPs for CBB and GSBB. Rows 1–3 correspond to the sample size $n = 120, n = 480$ and $n = 1920$, respectively

the optimal block length choices presented in Table 1 provide ACP very close or equal to nominal one. For $n = 120$ ACPs obtained with CBB are often higher than those for GSBB. This can be observed in the third columns of Figs. 1, 2, 3, 4, 5, 6, 7, 8, 9, 10. The curve of differences in ACP is more often positive than negative. This effect usually vanishes for higher values of n. Interestingly, sometimes ACP for CBB presents a strong periodic structure (see results for **M2**, **M4**, **M6**). The curve has local minima for b equal to integer multiple of d. This fact results in high variability in ACP values. Slight change in the chosen b value can provide a big difference in ACP. **M2** and **M4** where created by adding a periodic component to **M1** and **M5**, respectively. In **M6** a very strong periodic component was added and additionally the error term was also multiplied by it to check how this affects results. Since GSBB preserves the periodic structure of the original data ACP curves are much flatter and such variability is not observed. The optimal block length choices for CBB (Table 1) for $n = 1920$ are usually higher than for GSBB, but for smaller values of n such dependence no longer holds.

Values of ACP for the optimal block length choices (Table 1) are very comparable for CBB and GSBB independently on the chosen model when $n = 480$ or $n = 1920$. For $n = 120$ CBB often outperforms GSBB (see results for **M1**–**M3**, **M6** and **M9**). The highest difference can be found for **M2** when ACP for CBB is equal to 95.2 % and for GSBB 90.1 %.

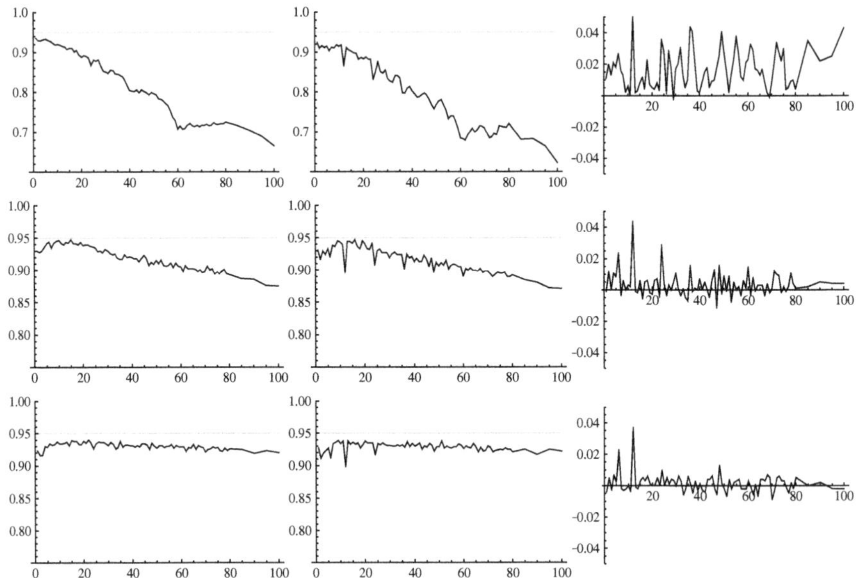

Fig. 9 First and second column: ACPs (*black lines*) for **M9** for $b \in \{1, 2, \ldots, 79, 80, 90, 100\}$ together with nominal coverage probability (*grey line*) for CBB and GSBB, respectively. Third column: differences between ACPs for CBB and GSBB. Rows 1–3 correspond to the sample size $n = 120$, $n = 480$ and $n = 1920$, respectively

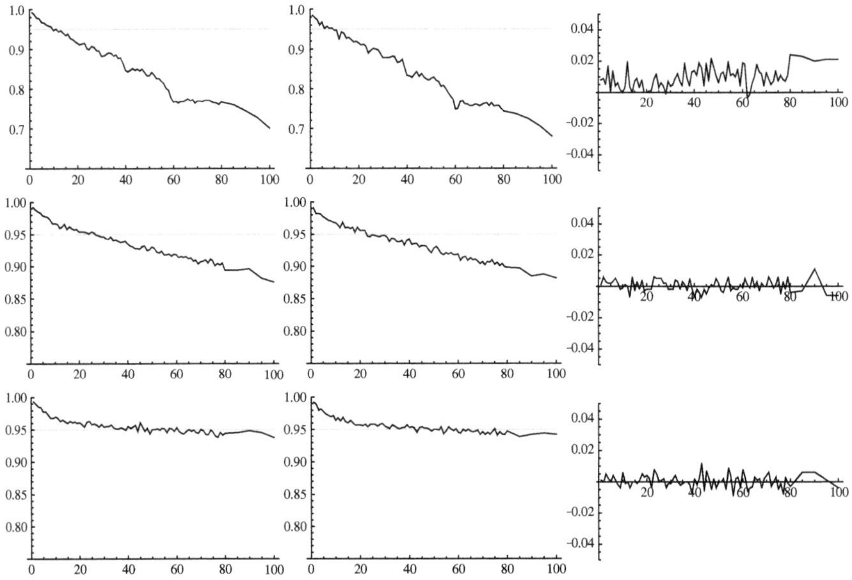

Fig. 10 First and second column: ACPs (*black lines*) for **M10** for $b \in \{1, 2, \ldots, 79, 80, 90, 100\}$ together with nominal coverage probability (*grey line*) for CBB and GSBB, respectively. Third column: differences between ACPs for CBB and GSBB. Rows 1–3 correspond to the sample size $n = 120$, $n = 480$ and $n = 1920$, respectively

Table 1 Optimal block length choice for **M1–M10** together with ACP for optimal b values

Model	n	CBB		GSBB	
		Optimal block length	ACP (%)	Optimal block length	ACP (%)
M1	120	7	91.0	16	89.7
	480	10, 11	94.2	20	94.5
	1920	63, 72	95.0	24	95.0
M2	120	18	95.2	7	90.1
	480	11, 43	95.0	17	94.0
	1920	51, 69	95.1	35, 37	95.0
M3	120	1, 5	93.6	1	91.7
	480	10, 15	94.8	19	95.3
	1920	20, 51	95.0	18, 19	95.0
M4	120	11	95.1	4	94.8
	480	12	95.1	17, 33	94.9
	1920	24	95.2	20, 29	95.0
M5	120	8	94.9	4	95.3
	480	23	95.0	8	95.0
	1920	58	95.0	11, 35, 53	95.0
M6	120	12	94.8	9	91.4
	480	16	94.4	10	94.4
	1920	52, 57	95.1	40	95.0
M7	120	4	93.8	6	93.2
	480	11, 16	94.7	12	95.0
	1920	3, 58	95.0	12	95.0
M8	120	15	85.5	13	85.9
	480	37	90.8	42	91.0
	1920	65	94.7	48	94.8
M9	120	1	93.5	1	92.4
	480	15	94.7	16	94.7
	1920	15	93.9	9	93.9
M10	120	11	95.0	6, 8	95.2
	480	24	95.1	25	95.3
	1920	50, 65, 70	95.0	48, 57, 62, 64	95.0

Rows refer to considered models. Columns 3, 4 and 5, 6 contain results for CBB and GSBB, respectively. For both methods optimal block lengths for three sample sizes $n = 120$, $n = 480$ and $n = 1920$ are presented

It seems that GSBB definitely can be used instead of CBB for longer datasets. For the shortest considered sample size CBB provides ACPs for optimal b choices closer to the nominal coverage level. On the other hand, sometimes a periodic variability in ACP curve for CBB can be observed. As a result small change in b value can provide a big change in coverage probability value, which is not the case for GSBB. Thus, the choice of the bootstrap method needs to be done very carefully for short samples, especially that so far no method of the optimal block length choice for the PC time series is known.

Acknowledgments Research of Anna Dudek was partially supported by the Polish Ministry of Science and Higher Education and AGH local grant.

References

Antoni J (2009) Cyclostationarity by examples. Mech Syst Sign Process 23(4):987–1036

Bloomfield P, Hurd HL, Lund R (1994) Periodic correlation in stratospheric ozone data. J Time Ser Anal 15:127–150

Bloomfield P, Hurd HL, Lund RB, Smith R (1995) Climatological time series with periodic correlation. J Clim 8:2787–2809

Chan V, Lahiri SN, Meeker WQ (2004) Block bootstrap estimation of the distribution of cumulative outdoor degradation. Technometrics 46:215–224

Dudek AE, Hurd H, Wójtowicz W (2013) perARMA: package for periodic time series analysis, R package version 1.5. http://cran.r-project.org/web/packages/perARMA

Dudek AE, Leśkow J, Politis D, Paparoditis E (2014) A generalized block bootstrap for seasonal time series. J Time Ser Anal. doi:10.1002/jtsa.12053. Accessed 27 NOV 2013

Hurd HL, Miamee AG (2007) Periodically correlated random sequences: spectral theory and practice. Wiley, Hoboken

Jones R, Brelsford W (1967) Time series with periodic structure. Biometrika 54:403–408

Leśkow J, Synowiecki R (2010) On bootstrapping periodic random arrays with increasing period. Metrika 71:253–279

Politis DN (2001) Resampling time series with seasonal components, in frontiers in data mining and bioinformatics. In: Proceedings of the 33rd symposium on the interface of computing science and statistics, Orange County, CA, 13–17 June 2001, pp 619–621

Politis DN, Romano JP (1992) A circular block-resampling procedure for stationary data. Wiley series in probability and mathematical statistics: probability and mathematical statistics. Wiley, New York, pp 263–270

Synowiecki R (2007) Consistency and application of moving block bootstrap for nonstationary time series with periodic and almost periodic structure. Bernoulli 13(4):1151–1178

Synowiecki R (2008) Metody resamplingowe w dziedzinie czasu dla niestacjonarnych szeregów czasowych o strukturze okresowej i prawie okresowej. PhD Thesis at the Departement of Applied Mathematics, AGH University of Science and Technology, Krakow, Poland. http://winntbg.bg.agh.edu.pl/rozprawy2/10012/full10012.pdf

Part II
Applications of Cyclostationarity

Modeling of Gear Transmissions Dynamics in Non-stationary Conditions

Fakher Chaari and Mohamed Haddar

Abstract Dynamic behavior of gear transmissions running under non-stationary operating conditions is extremely different from that operating in stationary conditions. The main feature that makes this difference is the variability of speed. Three main cases where it is possible to observe speed variation: start-up, shut down and time varying loading conditions. In this chapter, these three regimes will be discussed using dynamic modeling of gear transmission. Both amplitude and frequency modulations are observed in vibration signatures. The case study of wind turbine transmission is presented at the end of the chapter showing clearly this phenomenon.

1 Introduction

Gears are commonly used in several aerospace, automotive, and heavy industry mechanical systems. They are characterized by their ability to transmit large torques with an extreme amount of efficiency. However vibration levels, dynamic loads attenuation and control remain a key concern for researchers. Despite the improved nowadays technologies for design and manufacturing of gear systems, operating conditions such as variable speed and load, repetitive start-up and shut downs can be a source of malfunction giving rise to defects and undesirable noise and vibration. The so-called non-stationary regimes should be well characterized in order to help diagnosing gear systems in such conditions. Non-stationarity for a gear transmission is mainly related to speed variation. Start-up and shut down are the main typical examples. Critical dynamic loads may occur during these phases. If the starting load is higher than normal operating conditions, and the gear system is started and stopped frequently,

F. Chaari (✉) · M. Haddar
National School of Engineers of Sfax, Bp 1173, 3038 Sfax, Tunisia
e-mail: Fakher.chaari@gmail.com

M. Haddar
e-mail: mohamed.haddar@enis.rnu.tn

F. Chaari et al. (eds.), *Cyclostationarity: Theory and Methods*,
Lecture Notes in Mechanical Engineering, DOI: 10.1007/978-3-319-04187-2_8,
© Springer International Publishing Switzerland 2014

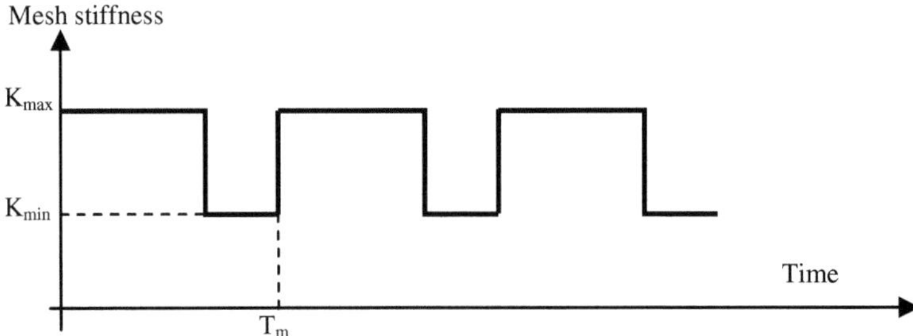

Fig. 1 Gear mesh stiffness evolution

failures can occur. Time varying loading conditions can be at the origin of fluctuating speeds. In fact, external load are often not constant. Real machines are subjected to real operating conditions, which generally are variable.

In the last decades, due to the rapid growth of computational power, novel modeling and simulation techniques were introduced. Since then, model based investigations for complex gearbox systems have been used very often. It is clear that, because of the cost of experiments and safety reasons, model based approach is the most feasible form in order to describe dynamic behavior of gearboxes in non-stationary conditions.

This chapter will give an overview about modeling techniques used to describe dynamic behavior of gear systems in non-stationary operating condition. Examples of start-up, shut down and time varying loading conditions will be discussed.

2 Overview on Gear Dynamics Modeling

Gearboxes are mainly used to provide speed and torque transform from a driving machine to a driven or connected mechanical device. Internal excitation source for gear system is caused by the time varying mesh stiffness considered as the main source of excitation of the system and at the origin of the observed noise and vibrations (Chaari et al. 2008, 2009). The evolution of the mesh stiffness is similar to rectangular pulse wave (Fig. 1). For a contact ratio less than 2, the maximum value of stiffness corresponds to the mesh of two teeth pairs whereas the minimum value corresponds to the mesh of only one pair (Chaari et al. 2006).

Fluctuation of the input velocity and torque caused by the motor at its transient regime and variable loading conditions are the main external excitation sources to a gear set. Sika et al. (2008) studied the backlash and backstrike effect and the influence of gear tooth geometry as well as shaft and gear-shaft-bearing casing positioning on the dynamic behavior of a gear system. Start-up of gear system powered by an electric motor was also investigated by Hugues (1993) and Khabou et al. (2011).

Defects can alter the dynamic behavior of gearboxes. They can be divided into three categories: manufacturing, assembly and running defects (Bartelmus et al. 2009b). Modeling of manufacturing errors such as profile or eccentricity errors can be achieved by adding a displacement function on the line of action (Bartelmus et al. 2010). Running errors which usually occur in teeth like cracks or breakage are modelled by a reduction in the mesh stiffness according to the severity of the defect (Chaari et al. 2009).

Bartelmus et al. (2010) studied the effect of time varying loading conditions on the dynamic behavior of a two stage gearbox and planetary gearbox in presence of distributed faults and shows that an increase in the vibration levels is observed when load increases. Bartelmus and Zimroz (2009a, b) noticed that variable load frequency and carrier rotational frequency are the main frequency components responsible of vibration modulations.

Walha et al. (2009) studied the effect of backlash on the dynamic response of a two stage gearbox. It was observed that tooth separation occurs at the transient regime with increase in the vibration level. Chaari et al. (2012) showed that load variation induces speed variation, which causes a variation in the gearmesh stiffness period. Bartelmus et al. (2010) presented models based on two mechanical systems used in the mining industry with original transmission error function expressing changes in technical condition and load variation. Khabou et al. (2011) studied the dynamic behavior of a single stage spur gear reducer in transient regime such as start up and acyclism.

3 Modeling of Non-stationary Operating Conditions

3.1 Start-Up of a Gear Transmission

Figure 2 shows a model of a single stage spur gear system. Eight degrees of freedom are considered (Khabou et al. 2011). The pinion has Z_1 and wheel have Z_2 teeth. Moments of inertia I_{11}, I_{12}, I_{21} and I_{22} corresponds respectively to driving motor, pinion, wheel and driven machine. The motor develops a driving torque C_m and the driven machine opposes a load C_L. The gear mesh stiffness $k_m(t)$ is modelled by linear spring acting on the line of action of the meshing teeth.
Transmission error can be expressed by (Velex 1988):

$$\delta(t) = (x_1 - x_2)\sin\alpha + (y_1 - y_2)\cos\alpha + \theta_{12}r_{b12} + \theta_{22}r_{b21}. \tag{1}$$

x_1, y_1 and x_2, y_2 are the translations of driving and driven shafts. $\theta_{11}, \theta_{12}, \theta_{21}$ and θ_{22} are the rotational degrees of freedom of driving motor, pinion, wheel and driven machine. α is the pressure angle. The base radius of the pinion and the wheel are respectively r_{b12} and r_{b21}.
The mesh period is expressed by

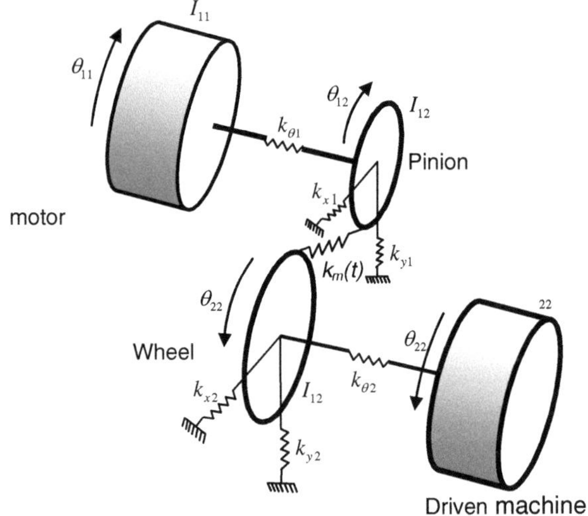

Fig. 2 Single stage spur gear system modeling (Khabou et al. 2011)

$$T_m = \frac{60}{N_1 Z_1}.$$

(2)

where N_1 is the rotational speed of motor.

Taking into account Lagrange formalism, the differential equation of motion of the adopted system is:

$$[M]\{\ddot{q}\} + [C]\{\dot{q}\} + [K(t)]\{q\} = \{F_{ext}(t)\}.$$

(3)

q is the vector of the degrees of freedom given by:

$$q = \{x_1, y_1, \theta_{11}, \theta_{12}, x_2, y_2, \theta_{21}, \theta_{22}\}^T$$

(4)

Expressions of mass matrices M, damping matrix C, the time varying stiffness matrix K(t) and the external applied forces vector can be found in Khabou et al. (2011)

During the start-up phase of an electric motor driving the transmission, the transient regime shown by zone (A) in Fig. 3 is characterized by an increasing rotational speed. This regime is followed immediately by the steady state regime (zone B) with nominal rotational speed ω_n of the motor.

Figure 4 shows the evolution of the mesh stiffness during the transient regime with non-periodic evolution explained by the fact that the speed is growing from zero to nominal speed.

Figure 5 shows the evolution of the transmission error computed after solving the equation of motion (3).

It is well observed that start-up is characterized by high vibration level explained by the fact that the system is excited at its eigenfrequencies. A more stable regime follows with lower vibration levels.

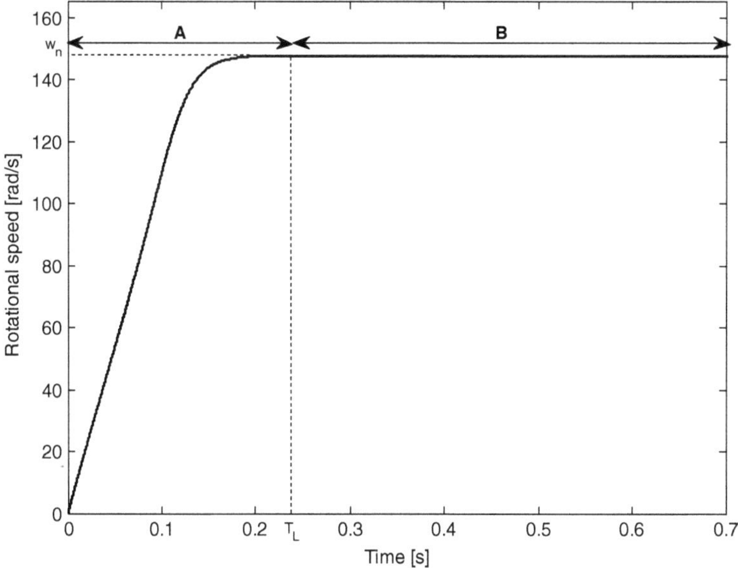

Fig. 3 Rotational speed evolution of induction motor (Khabou et al. 2011)

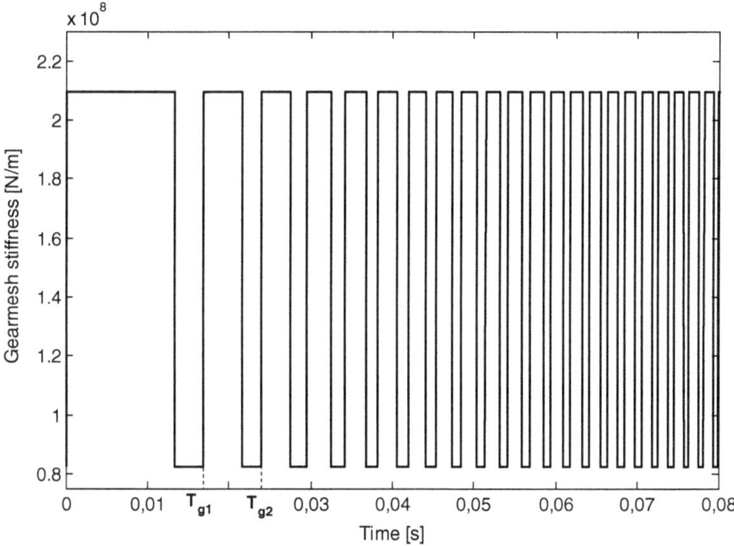

Fig. 4 Evolution of mesh stiffness during start-up (Khabou et al. 2011)

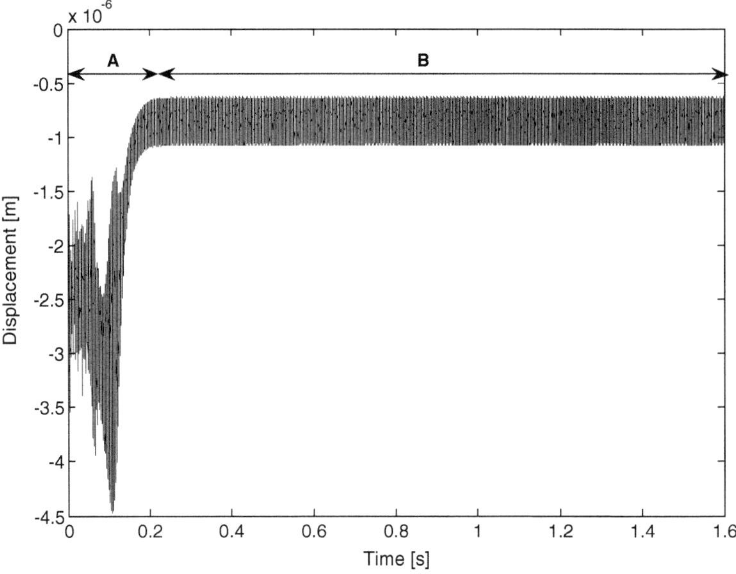

Fig. 5 Evolution of the transmission error (Khabou et al. 2011)

3.2 Shut Down of a Gear Transmission

In this case, the single stage spur gear system is connected to a disc brake system in order to stop the system at a desired instant (Fig. 6). The expression of the braking torque in this case is given by:

$$C_b = \frac{4\mu N\beta}{3\sin(\frac{\beta}{2})}\left(\frac{R_1 + R_2}{2}\right)\left[1 - \frac{R_1 R_2}{(R_1 + R_2)^2}\right] \tag{5}$$

where R_1, R_2, β are the geometrical characteristics of the brake, N is the normal force applied on the pads and μ is the friction coefficient.

The stop phase will be characterized by a fall in the speed until reaching 0. So, mesh stiffness evolution will be characterized by an increase of its period following the decrease of the speed of the gear set.

Figure 7 shows the evolution of the transmission error.

We note maximum amplitude of the error at the braking moment. The introduction of the sudden braking moment is at the origin of this amplitude amplification.

3.3 Time Varying Loading Conditions

The torque developed by a synchronous induction motor varies with its speed when it accelerates to reach its nominal operating speed (Wright 2005).

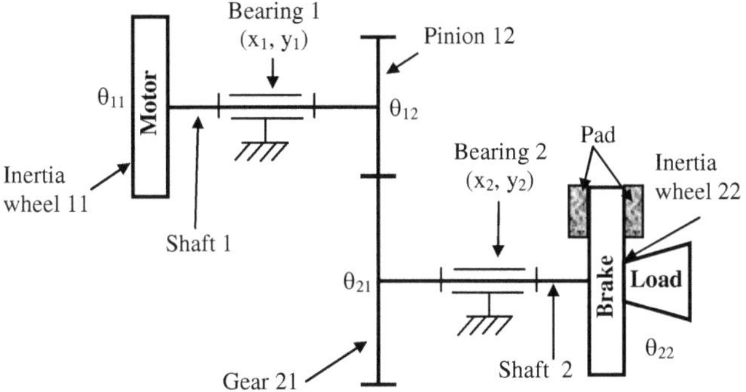

Fig. 6 Model of gear system including brake (Khabou et al. 2014)

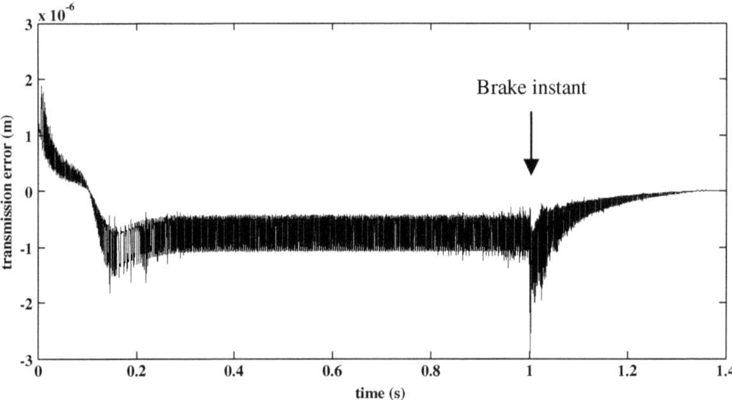

Fig. 7 Evolution of the transmission error in the case of braking (Khabou et al. 2014)

Let us consider an asynchronous motor driving the gearbox presented in Sect. 3.1.

When energized, the motor torque C_m must always exceed the torque absorbed by the load C_L since the excess torque $C_{net} = C_m - C_L$ is necessary to accelerate the system to its steady running speed for which $C_{net} = C_m = C_L$ (Fig. 8).

The motor driving torque for a squirrel cage electric motor can be expressed by:

$$C_M = \frac{C_b}{\left(1 + (s_b - s)^2 \left(\frac{a}{s} - bs^2\right)\right)} \tag{6}$$

s_b and C_b are the slip and torque at break-down (at maximum torque), a and b are all constant properties of the motor and s is the proportional drop in speed given by:

$$s = (N_s - n_r)/Ns \tag{7}$$

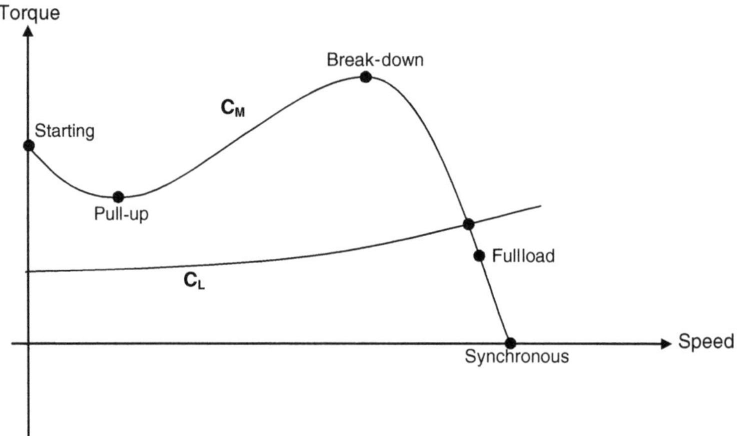

Fig. 8 Torque-speed characteristics for the motor and load

where Ns is the synchronous speed and n_r is the motor rotational speed. As the load torque increase the motor should provide the necessary torque to continue driving the mechanism but the speed will be decreased following the characteristic given in Fig. 8.

If we consider a load applied to the transmission fluctuating in a saw-tooth shape as presented in Fig. 9 and in order to continue transmitting power, the rotational speed of the motor will change which will induce a variation in the gear mesh stiffness period. If load increase, speed decrease and mesh period increase and vice versa.

Figure 10 shows the acceleration simulated on the pinion bearing and Fig. 11 shows its spectrum.

High amplitude modulation in time responses is observed. Amplitudes of acceleration signal increases when load increases and vice versa. Spectrum shows multiple sidebands around mesh frequency and harmonics. Such results may lead to erroneous diagnosis of the gear transmission (it can be seen as tooth local fault), if loading conditions are not well identified. So for non-stationary operating conditions separated time and frequency analysis is not suited for dynamic analysis and diagnosis.

Joint time frequency analysis can be a good alternative to characterize frequency content and localize speed variation caused by load variation. It consists in a three-dimensional time, frequency and amplitude representation of a signal very useful to detect transient behavior in a signal. Short-time Fourier transform (STFT) is the most popular time frequency representation defined by for a given time signal x(t) as:

$$\text{STFT}(t, \omega) = \int_{-\infty}^{+\infty} x(t) w(t - \tau) e^{-j\omega t} dt \tag{8}$$

where w(t) is the window function, commonly a Hann window or Gaussian "hill" centered around zero.

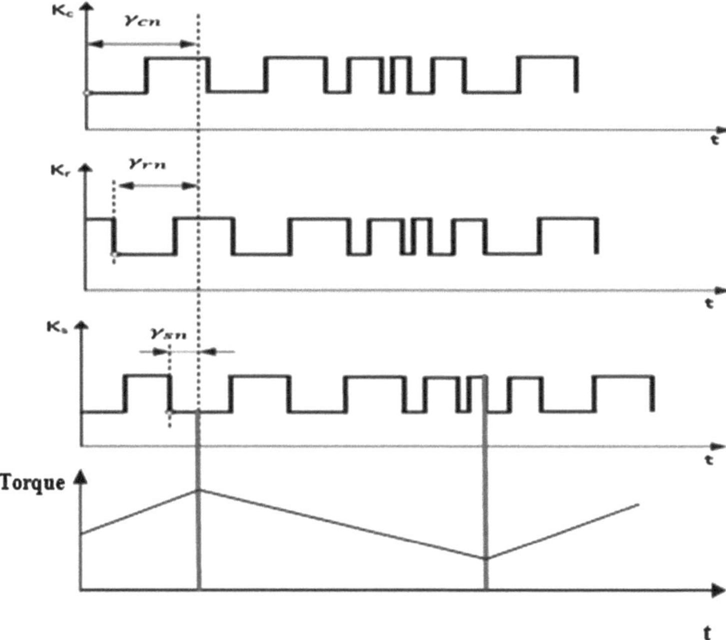

Fig. 9 Evolution of load and relation to gearmesh stiffness function (Chaari et al. 2013)

STFT of the acceleration on pinion is presented in Fig. 12.

From Fig. 12 one can note a sawtooth shapes of mesh frequency evolution and its harmonics with a same period of load. Mesh frequency harmonics vary from a minimum value (corresponding to the maximum applied load) to a maximum value (corresponding to the minimum applied load). Dark red in mesh frequency fluctuation indicates the vicinity of some eigenfrequencies. So, sidebands noticed in spectral analysis are caused by non-stationary operating condition and not by the presence of a defect.

Several experimental works showed similar evolutions (Randall 1982; Bartelmus 1992, 2001; Bartelmus et al. 2009a). In fact, both amplitude and frequency modulation exist in the signal if gearbox works under time varying operating conditions which makes diagnosis process complicate if loading conditions are not well identified.

4 Case Study: Wind Turbine Modeling

Wind turbines are considered as one of the most promising mechanical devices to produce so called renewable energy. Transmission systems used in such devices are subjected to very hard operating conditions depending mainly on the wind variability.

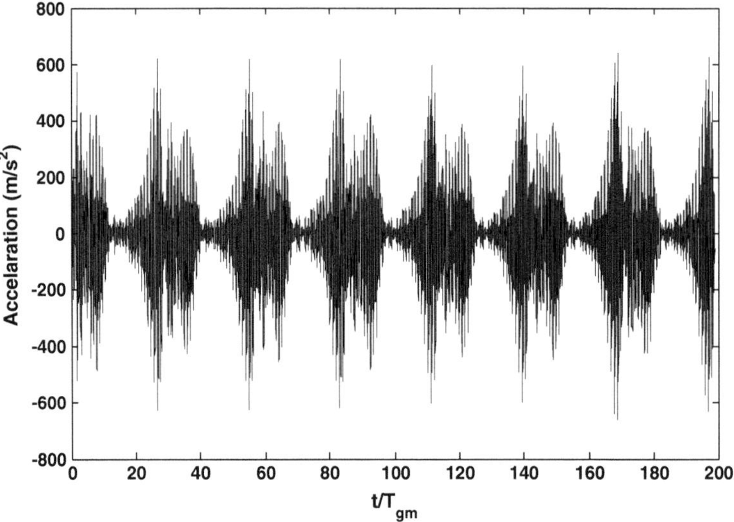

Fig. 10 Acceleration on pinion bearing (Chaari et al. 2012)

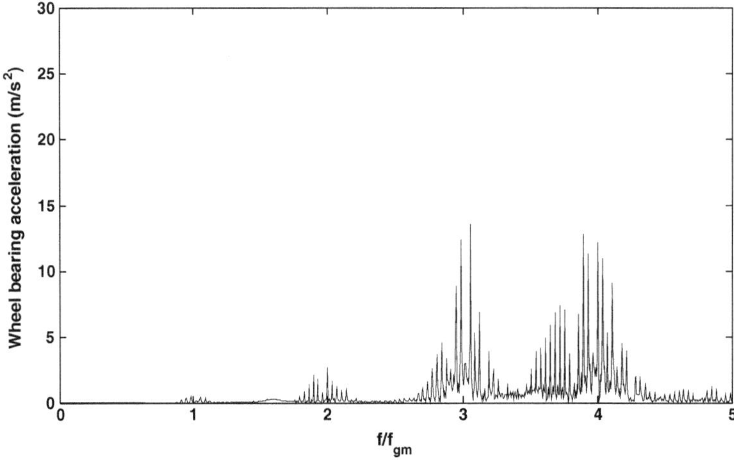

Fig. 11 Acceleration on pinion bearing for variable loading conditions *a- time response –b-spectrum* (Chaari et al. 2012)

Experience coming from commercial monitoring systems and recent communication regarding their reliability show that there is a need to improve both design, and diagnostics techniques used in order to extend wind turbine lifetime. For wind turbines, model/characteristics of input is unknown and rather stochastic than deterministic, gearbox systems are working as multiplicators (they increase speed and decrease torque) and electric generator is used as receiving machine. These differences are crucial for dynamics of transmissions and their degradation processes.

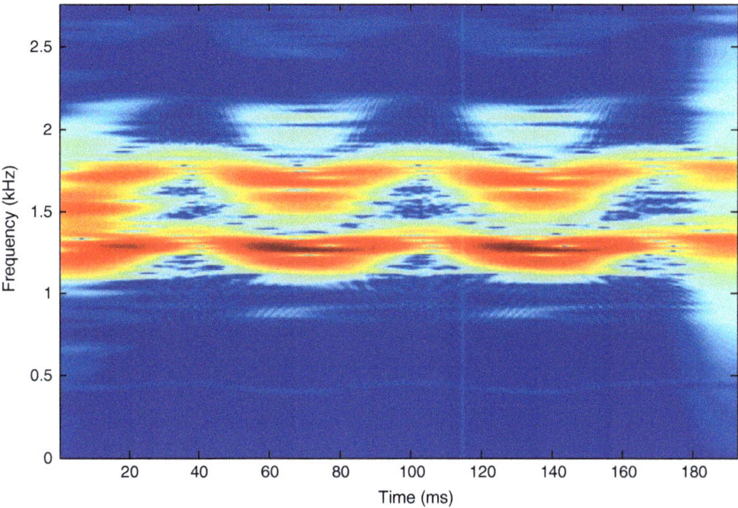

Fig. 12 STFT of the acceleration on pinion bearing for variable loading conditions (Chaari et al. 2012)

Wind turbine system can be divided into three subsystems

- System converting wind to input torque (including blades and input shaft)
- Gear transmission to multiply rotational speed and reduce torque. Two solutions can be adopted : three stages parallel axis gear or three stages gear system including planetary gear
- receiving machines i.e. asynchronous electric power generator

As a case study we will take the example of a wind turbine including three stages gearbox. The first stage is a planetary gearbox (PG). The second and third ones are parallel shafts gearboxes. The input component which is directly connected to the turbine is the carrier of the PG. The output component connected to the generator is the wheel of the third stage gearbox (Fig. 13). For the first stage, the sun (s), the ring (r), the carrier (c) and the N planets (p) are considered as rigid bodies. The bearings are modelled by linear springs. Gearmesh stiffness is modelled by linear springs acting on the lines of action. Each component has three degrees of freedom, two translations x_i, y_j and one rotation w_j, where $w_j = r_j\,\theta_j$ ($j = c, r, s, 1, \ldots .N$), r_j is the base radius of the gears (sun, ring, planet) and the distance between the centre of the sun and the planets' rotation centre. Damping is introduced as modal viscous damping. Radial and tangential coordinates x_p, y_p describe planet deflections and e_{sp} represents respectively the transmission error on the sun-planets gearmesh and on the planets-ring gearmesh, induced by wear. Translations are measured relative to the frame $(O, \vec{i}, \vec{j}, \vec{k})$ fixed to the carrier and rotating with constant angular speed Ω_C relative to a stationary reference frame. Circumferential planet positions are specified by fixed angles φ_p measured relative to the rotating frame with $\varphi_1 = 0$.

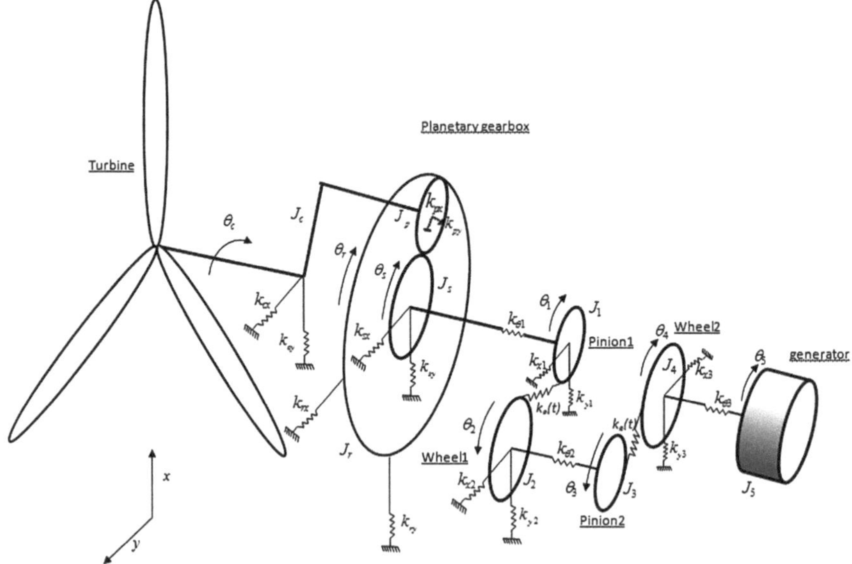

Fig. 13 Model of 3 stage transmission in wind turbine

k_{jx}, k_{jy}, k_{jw} represent respectively the bearing stiffness along x and y directions and k_{jw} are the torsional stiffness (j = c, r, s). The three degrees of freedom k_{px}, k_{py} are the bearing stiffness of each planet.

The first stage is connected to the second stage by a shaft modelled by a torsional spring $k_{\theta 1}$ and a damping c_1. The second stage is modelled by 6 DOF which are the translations of pinion 1(x_1, y_1) and wheel 1 (x_2, y_2) along x (horizontal) and y (vertical) directions and the rotations (θ_1, θ_2) of pinion1 and pinion 2.

The second stage is connected to third stage by a shaft modelled by a torsional spring $k_{\theta 2}$. The third stage is modelled by 6 DOF which are the translations of pinion 2 (x_3, y_3) and wheel 2 (x_4, y_4) along x (horizontal) and y (vertical) directions and the rotations (θ_3, θ_4) of pinion 2 and wheel 2.

The third stage is connected to the generator by a shaft modelled by the torsional stiffness $k_{\theta 2}$. To the generator is assigned the DOF θ_4.

Bearings supporting the second stage are modelled by linear springs k_{x1} and k_{y1} acting along x and y axis. Bearings supporting intermediate shaft are modelled by linear springs k_{x2} and k_{y2}. Bearings supporting third stage are modelled by linear spring k_{x3} and k_{y3}.

The time varying mesh stiffness is chosen to be a step function. In the first stage, two kind of mesh stiffness are considered: stiffness between sun and each planet-p $k_{sp}(t)$ and stiffness between each planet-p with the ring $k_{rp}(t)$. Two mesh stiffness functions $k_e(t)$ are introduced between pinion 1 and wheel 2 (second stage) and between pinion 2 and wheel 2 (third stage).

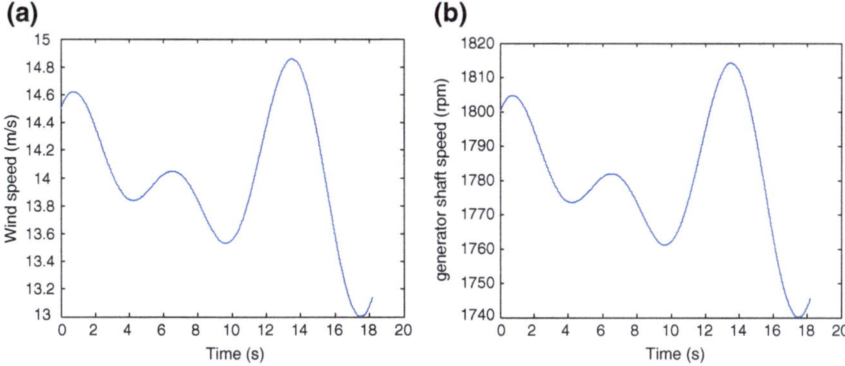

Fig. 14 Evolution of wind speed (**a**) and the output speed on the generator (**b**)

The following inertia are considered: J_c for the carrier, Js for the sun, Jr for the ring Jp for each planet, J_1 for the pinion 1, J_2 for wheel, J_3 for pinion 2, J_4 for wheel 2 and J_5 for the generator.

The system has $21 + 3 * N$ degrees of freedom (DOF).

An input torque C_m is applied to the carrier resulting from the produced torque on turbine. A resisting torque C_r is opposed by the load represented by the generator. Damping is introduced in parallel of each stiffness.

In order to compute the input torque and relate it to input turbine speed we have to know the power produced by the wind. It can be expressed as (DNV/RISO 2002):

$$P = 0.5 \ \rho \ V^3 A \ C_p \tag{9}$$

where: P is the output power, ρ is the air density, V is the free wind speed A is the rotor area and Cp is the efficiency factor

Starting from a known wind speed evolution presented in Fig. 14, it is possible to compute the output power. If it is divided by the instantaneous angular velocity of the turbine, the input torque C_m can be obtained.

Introducing the speed evolution and the input torque and computing the different mesh stiffness functions allows obtaining the general equation of motion to solve.

Figure 15a shows the time response on the input stage bearing i.e. the carrier bearing and Fig. 15b shows the response on output shaft bearing.

It is well observed the modulation on signals induced by the change of speed and torque developed on the turbine.

In order to better understand the frequency content evolution, it is necessary to compute a time frequency analysis. Figure 16 presents the STFT of time response given in Fig. 15a.

It is well observed the time evolution of the mesh frequency harmonics which indicate a frequency modulation in addition to the amplitude modulation observed in time responses. Such result was obtained experimentally

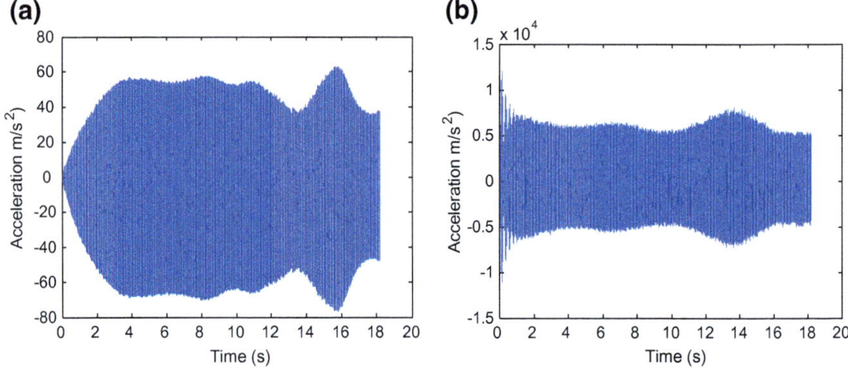

Fig. 15 Dynamic response on the bearings of input (**a**) and output shaft (**b**)

Fig. 16 STFT of the time response on the on the input shaft

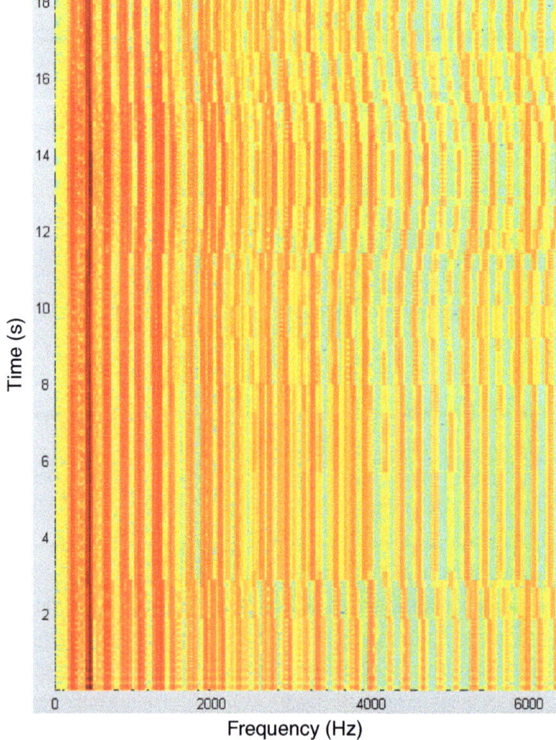

Similar results was obtained from experimental data measured on wind turbine by Zimroz et al. (2011) shown by using STFT that the vibration signal from the wind turbine gearbox is non-stationary in terms of amplitude and frequency content (Fig. 17). He noticed impulsive disturbance that causes both vibration and speed

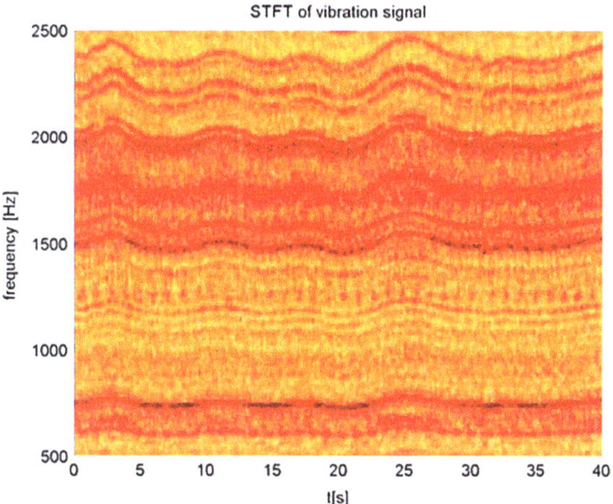

Fig. 17 Time Frequency representation of wind turbine gearbox vibration according to Zimroz et al. (2011)

profile variable concluding that time-varying operating conditions influence both the frequency content and amplitudes of harmonics.

5 Conclusion

In this chapter, the dynamic behavior of gear transmission in non-stationary conditions is investigated. The main excitation of the transmission is the mesh stiffness fluctuation which is considered as the source of the observed noise and vibration. Three cases of non-stationary operating conditions were studied: start-up, shut down and time varying loading conditions where speed variation is observed. Dynamic behavior showed a deep amplitude and frequency modulation. Spectral analysis is unable to detect speed variation and can lead to an erroneous diagnosis of the transmission. Time frequency techniques should be used in this case in order to describe the frequency content variation. The case study on wind turbine gear transmission dynamics confirms this fact. For a gear transmission and in order to diagnose its health, it is than necessary to take into not also eventual defects but also operating conditions.

References

Bartelmus W (1992) Vibration condition monitoring of gearboxes. Mach Vib 1:178–189
Bartelmus W (2001) Mathematical modelling and computer simulations as an aid to gearbox diagnostics. Mech Sys Sig Proc 15:855–871

Bartelmus W, Zimroz Z (2009a) Vibration condition monitoring of planetary gearbox under varying external load. Mech Syst Sig Proc 23:246–257

Bartelmus W, Zimroz Z (2009b) A new feature for monitoring the condition of gearboxes in non-stationary operating conditions. Mech Syst Sig Proc 23:1528–1534

Bartelmus W, Chaari F, Zimroz R, Haddar M (2010) Modelling of gearbox dynamics under time varying non-stationary operation for distributed fault detection and diagnosis. Eur J Mec A/Sol 29:637–646

Chaari F, Fakhfakh T, Haddar M (2006) Simulation numérique du comportement dynamique d'une transmission par engrenages en présence de défauts de denture. Méc Ind 6:625–633

Chaari F, Baccar W, Haddar M (2008) Effect of spalling or tooth breakage on gearmesh stiffness and dynamic response of a one-stage spur gear transmission. Eur J Mec A/Sol 27:691–705

Chaari F, Fakhfakh T, Haddar M (2009) Analytical modelling of spur gear tooth crack and influence on gearmesh stiffness. Eur J Mec A/ Sol 27:461–468

Chaari F, Bartelmus W, Zimroz R, Fakhfakh T, Haddar M (2012) Gearbox vibration signal amplitude and frequency modulation. Sh Vib 19:635–652

Chaari F, Abbes MS, Rueda FV, del Rincon AF, Haddar M (2013) Analysis of planetary gear transmission in non-stationary operations. Front Mec Eng 8:88–94

DNV, RISO(2002) Guidelines for design of wind turbines, 2nd edn. Det Norske Veritas, Copenhagen

Hugues JH (1993) Contribution à l'étude dynamique lors du démarrage de chaines cinématiques à engrenages entraînés par moteur éléctrique, PhD, INSA Lyon, France

Khabou MT, Ksentini O, Jarraya A, Abbes MS, Chaari F, Haddar M (2014) Influence of disk brake friction on the dynamic behaviour of a directly coupled spur gear transmission. Mult Mod Mat Str 10 (in press)

Khabou MT, Bouchaala N, Chaari F, Fakhfakh T, Haddar M (2011) Study of a Spur Gear Dynamic Behavior in Transient Regime. Mec Sys Sig Pro 25:3089–3101

Randall RB (1982) A new method of modelling gear faults. J Mech Des 104:259–267

Sika G, Velex P (2008) Analytical and numerical analysis of gears in the presence of engine acyclism. ASME J Mech Des 130:1–6

Velex P (1988) Contribution à l'analyse du comportement dynamique de Réducteurs à engrenages à axes parallèles. PhD INSA Lyon, France

Walha L, Fakhfakh T, Haddar M (2009) Nonlinear dynamics of a two-stage gear system with mesh stiffness fluctuation, bearing flexibility and backlash. Mech Mach Th 44:1058–1069

Wright D (2005) Class Notes on Design and Analysis of Machine Elements. The University of Western Australia, Department of Mechanical and Materials Engineering, Crawley, Perth, Australia

Zimroz R, Urbanek J, Barszcz T, Bartelmus W, Millioz F, Martin N (2011) Measurement of instantaneous shaft speed by advanced vibration signal processing—application to wind turbine gearbox. Met Meas Syst 18:701–712

Effects of Satellite Motion on the Received Signal in GPS

Antonio Napolitano and Ivana Perna

Abstract In this chapter, the effects of the relative motion between a GPS satellite and a stationary receiver on the Earth are addressed. An analysis of the satellite motion is carried out to justify the assumption of constant relative radial speed within observation intervals adopted in the applications. It is shown that the transmitted cyclostationary signal is still cyclostationary at the receiver but with different cycle frequencies and cyclic features. Moreover, the transmitted and received signals are not jointly cyclostationary but, rather, jointly spectrally correlated. The implications of this statistical characterization on synchronization and parameter estimation problems are discussed.

1 Introduction

The Global Positioning System (GPS) is one of the most popular Global Navigation Satellite Systems (GNSS). GNSS are an essential tool for positioning and navigation in many key sectors, including aeronautics, ground and sea transportation, and infrastructure monitoring. In several of these sectors, a challenging problem is counteract the effects of disturbance signals that produce significant performance degradation.

In the last two decades, cyclostationarity properties of signals have been successfully adopted to significantly improve performances of signal processing algorithms due to their capability to counteract the effects of noise and interference. In particular, since almost-all modulated signals adopted in communications, radar, sonar, and telemetry are cyclostationary or, more generally, almost-cyclostationary,

A. Napolitano (✉) · I. Perna
University of Napoli "Parthenope", Department of Engineering, Napoli, NA, Italy
e-mail: antonio.napolitano@uniparthenope.it

I. Perna
e-mail: ivana.perna@uniparthenope.it

F. Chaari et al. (eds.), *Cyclostationarity: Theory and Methods*,
Lecture Notes in Mechanical Engineering, DOI: 10.1007/978-3-319-04187-2_9,

cyclostationarity-based algorithms have been exploited in weak-signal detection, minimum mean-squared error (MMSE) filtering, parameter estimation, synchronization, and source location (Gardner et al. 2006). The gain in performance with respect to classical algorithms based on a wide-sense stationary model for signals is due to an accurate characterization of the signal nonstationarity. Since the signal transmitted by a GPS satellite is cyclostationary (Napolitano and Perna 2013a), the advantages of cyclostationarity-based techniques can be potentially adopted in GPS receivers, for example, to counteract the effects of unintentional and intentional jammer, provided that the received signal is in turn cyclostationary and jointly cyclostationary with the transmitted one.

The relative motion between transmitter (TX) and receiver (RX) modifies the nonstationarity kind of the transmitted signal since it experiences a time-varying delay. In particular, several Doppler channels encountered in practice, modify the (almost-)cyclostationarity properties of the transmitted signal into more general kind of nonstationarity (Napolitano 2012, Chap. 7). Moreover, also the joint (almost-)cyclostationarity of transmitted and received signals is lost.

In the case of constant relative radial speed between TX, RX, and/or surrounding scatterers, the joint almost-cyclostationarity property of transmitted and received signals is preserved provided that the so-called narrow-band condition is satisfied, that is, provided that the product of signal bandwidth and data-record length is much smaller than the ratio of the medium propagation speed and the radial speed (Napolitano 2012, Sect. 7.5; Van Trees 1971, pp. 339–340).

The necessity to satisfy the narrow-band condition to preserve a useful and powerful mathematical model for the received signal and for the joint characterization of transmitted and received signals puts a limit on the maximum data-record length that can be adopted in cyclostationarity-based signal processing algorithms. This limit, in turn, puts a limit on the minimum signal-to-noise ratio (SNR) and signal-to-interference ratio (SIR) for which satisfactory performance can be achieved.

In order to overcome such a lower bound for SNR and SIR, larger data-record lengths should be adopted, leading to scenarios where the narrow-band condition is not satisfied. In such a case, the Doppler effect cannot be modeled as a simple frequency shift of the carrier and higher fidelity models should be considered for the received signals. In (Napolitano 2012, Sects. 4.2.4, 7.7) it is shown that the spectrally correlated processes are an appropriate model in several cases of interest.

In this chapter, the effects of satellite motion on the received signal in GPS are analyzed. A simplified analysis is carried out assuming a circular orbit for the satellite. This simplifying assumption on the orbit, however, will not lead to a simplified model for the received signal. In fact, in order to specify the model for the received signal, what is necessary is just to quantify the order of magnitude of the relative radial speed between satellite and receiver and the width of time intervals such that the radial speed can be assumed constant. The order of magnitude of the radial speed does not vary if an elliptic orbit instead of a circular one is considered. Similarly, corrections to the orbit due to relativistic effects do not involve rapidly time varying phenomena and can be neglected. Thus, time intervals where the radial speed of the circular orbit can be assumed constant are also such that the effective radial speed can be assumed constant.

In this chapter, it is shown that the relative radial speed between satellite and ground receiver can be considered constant in time intervals of interest in the applications, even when such time intervals assume their maximum values as in the case of strong contaminating jammer or indoor applications (Seco-Granados et al. 2012).

For a single-path Doppler channel, the received GPS signal is shown to be still cyclostationary, but with different cycle frequencies and cyclic features (cyclic auto-correlation functions and cyclic spectra) with respect to those of the transmitted one. In contrast, the transmitted and received signals are, in general, jointly spectrally correlated and they can be modeled as jointly cyclostationary only if the data-record length is such that the narrow-band condition is satisfied.

The derived analytical model for the cyclic statistics of the received signal and of the cross statistical characterization of transmitted and received signals is not influenced by the simplifying assumptions made on the satellite orbit. In contrast, the values of the parameters involved in the model depend of the effective orbit of the satellite. Thus, the derived model turns out to be useful when these parameters are estimated starting from the received signal.

Starting from the model of received signal developed in this chapter, in (Napolitano and Perna 2013b) a synchronization algorithm is presented that significantly outperforms the classical one based on the maximization of the magnitude of the narrow-band cross ambiguity function when the data-record length is augmented in order to counteract noise and interference.

The chapter is organized as follows. In Sect. 2, a simplified model for the satellite orbit is presented and the resulting propagation channel is derived in Sect. 3. In Sect. 4, the transmitted signal is described. The statistical characterization of the received signal is derived in Sect. 5. Conclusions are drawn in Sect. 6.

2 Satellite Orbit

Consider an orthogonal coordinate system whose origin O is at the center of the Earth and whose z axis coincides with the oriented line from South pole to North pole. Such a system is referred to as Earth-fixed Earth-centered (EFEC). It can be made inertial by freezing the reference frame at the instant of time when the signal acquisition process is started. Let $S(t)$ denote the satellite position in this system and $r(t)$ be the time-varying vector $S(t) - O$. Under the assumption of Earth with spherical shape and uniform mass density, the gravitational force acting on the satellite in S is given by the Newton's universal law of gravitation

$$F = -G \frac{mM}{r^3} r \tag{1}$$

where $G = 6.672 \times 10^{-11} \mathrm{m}^3\,\mathrm{kg}^{-1}\mathrm{s}^{-2}$ is the universal gravitation constant, $M = 5.974 \times 10^{24}\,\mathrm{kg}$ is the mass of the Earth, m is the satellite mass, and $r = \|\boldsymbol{r}\|$. Thus, the second law of dynamic $m\,(\mathrm{d}^2\boldsymbol{r}/\mathrm{d}t^2) = \boldsymbol{F}$ leads to

$$\frac{\mathrm{d}^2\boldsymbol{r}}{\mathrm{d}t^2} = -\frac{\mu}{r^3}\,\boldsymbol{r} \tag{2}$$

where $\mu \triangleq GM = 3.986 \times 10^{14}\,\mathrm{m}^3\mathrm{s}^{-2}$.

According to Kepler laws, the orbit of a satellite is an elliptic trajectory within a plane. The ellipse has a maximum extension at the apogee and a minimum at the perigee. The satellite moves more slowly in its trajectory as the distance from the Earth increases. A more accurate model for the satellite's orbit should account for the fact that the Earth is not spherical and has not a uniform distribution of mass and, moreover, there are additional forces and perturbations acting on the satellite.

In the following, the satellite orbit is modeled as circular and the satellite motion as uniform circular. This simplified kinematic model will not influence the accuracy of the model for the received signal. In fact, as shown in Sects. 3 and 5, in order to properly model the received signal, what is important is not an accurate description of the relative radial speed and acceleration laws, but, rather, their order of magnitude and the order of magnitude of their variations with time within observation intervals of interest in the applications. Of course, the values of the parameters in the model of the received signal will depend on the parameters of the kinematic model. However, this is not a problem in applications where such parameters are estimated.

For a uniform circular motion with radius r and (constant) angular speed ω_0, the magnitudes of the velocity vector $\mathrm{d}\boldsymbol{r}/\mathrm{d}t$ (tangent to the circular trajectory) and of the acceleration vector $\mathrm{d}^2\boldsymbol{r}/\mathrm{d}t^2$ are

$$v = \omega_0\,r \qquad a = \omega_0^2\,r \tag{3}$$

respectively. Accounting for (2), we have $a = \omega_0^2\,r = \mu/r^2$. Thus,

$$v = \omega_0\,r = \sqrt{\frac{\mu}{r}} \qquad \omega_0 = \frac{v}{r} = \sqrt{\frac{\mu}{r^3}} \tag{4}$$

and the satellite revolution period is

$$T_{\mathrm{sat}} = \frac{2\pi}{\omega_0} = 2\pi\sqrt{\frac{r^3}{\mu}}. \tag{5}$$

Let $R_E = 6378.13\,\mathrm{km}$ be the average Earth radius and h the satellite altitude with respect to the Earth's surface. Thus, $r = R_E + h = \mathrm{constant}$ for a circular orbit (Fig. 1). From (5) it follows that the satellite revolution period depends only on r, that is, on the altitude h of the satellite with respect to the Earth surface.

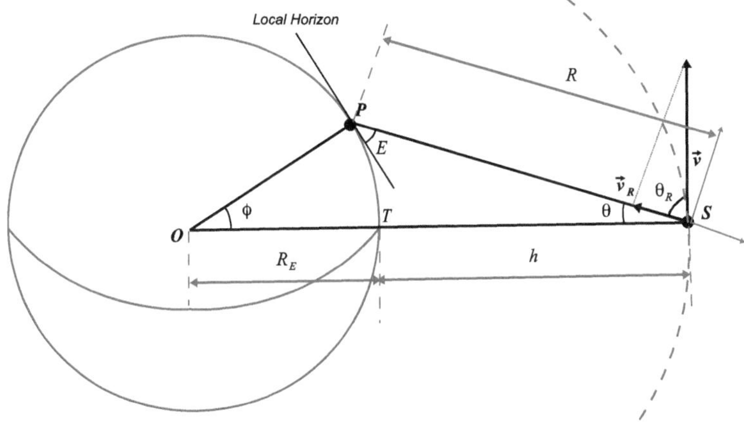

Fig. 1 Velocity vector and satellite radial speed in the Earth–satellite geometry

For a GPS satellite a typical value for the altitude is $h = 20182$ km. Consequently, $v = \sqrt{\mu/r} = 3.874$ km s^{-1} and $T_{\text{sat}} \simeq 11$h 58 m, that is, one half sidereal day.

Let $\boldsymbol{P}$ a *stationary point* on the Earth surface where the receiver is located. Since r is constant, the satellite position can be univocally determined by two of the following angles (Fig. 1) (Maral and Bousquet 2009, Sect. 2.1.6):

- *Elevation angle $E(t)$*, that is, the angle between the local horizon at the fixed point $\boldsymbol{P}$ and the satellite position vector $\boldsymbol{S}(t)$ measured in the plane containing the point $\boldsymbol{P}$, the Earth center $\boldsymbol{O}$, and the satellite $\boldsymbol{S}(t)$. For an *ascending satellite*, $E(t)$ ranges from 0 to $\pi/2$ as t increases: $\dot{E}(t) > 0$, where $\dot{E}$ denotes time derivative of E; for a *descending satellite*, $E(t)$ ranges from $\pi/2$ to 0 as t increases: $\dot{E}(t) < 0$.
- *Nadir angle $\theta(t) \in (0, \pi/2)$*, that is, the angle at the satellite between the direction $\boldsymbol{r}(t)$ of the satellite with respect to the Earth center $\boldsymbol{O}$ and the direction $\boldsymbol{S} - \boldsymbol{P}$.
- *Earth central angle $\phi(t)$*. For an *ascending satellite*, $\phi(t)$ ranges from $\pi/2$ to 0 as t increases: $\phi(t) = -\omega_0 t$; for a *descending satellite*, $\phi(t)$ ranges from 0 to $\pi/2$ as t increases: $\phi(t) = \omega_0 t$.

Only two of these angles are sufficient since (Fig. 2)

$$E(t) + \phi(t) + \theta(t) = \frac{\pi}{2} \tag{6}$$

In addition, from the Earth–satellite geometry (Fig. 2) we have

$$R(t) \cos E(t) = r \, \sin \phi(t) = \|\boldsymbol{S}(t) - \boldsymbol{P}'(t)\| \tag{7}$$

$$\tan E(t) = \frac{\|\boldsymbol{P} - \boldsymbol{P}'(t)\|}{\|\boldsymbol{S}(t) - \boldsymbol{P}'(t)\|} = \frac{r \, \cos \phi(t) - R_E}{r \, \sin \phi(t)} = \frac{\cos \phi(t) - \rho}{\sin \phi(t)} \tag{8}$$

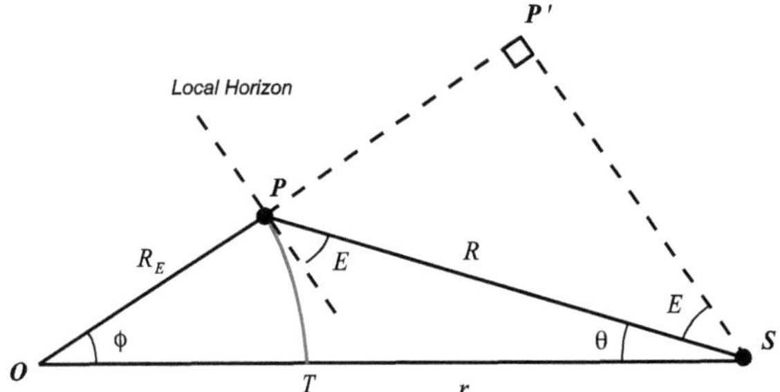

Fig. 2 Earth–satellite geometry

$$R(t) = \|S(t) - P\| = \frac{\|S(t) - P'\|}{\cos E(t)} = (R_E + h)\,\frac{\sin \phi(t)}{\cos E(t)} \tag{9}$$

where $\rho \triangleq R_E/r$. Therefore, under the assumption of uniform circular motion and descending satellite we have

$$\phi(t) = \omega_0\, t \tag{10}$$

$$E(t) = \tan^{-1}\left(\frac{\cos(\omega_0\, t) - \rho}{\sin(\omega_0\, t)}\right) \tag{11}$$

Let $\theta_R(t)$ be the angle between the velocity vector (tangent to the circular trajectory) in $S(t)$ and the direction $S(t) - P$ (Fig. 1). Using (6) it results

$$\theta_R(t) = \frac{\pi}{2} - \theta(t) = E(t) + \phi(t) \tag{12}$$

and the *relative radial speed* of the satellite $S(t)$ with respect to the receiver in P is given by

$$v_R(t) = v\,\cos\theta_R(t) = v\,\cos\left[\omega_0\, t + \tan^{-1}\left(\frac{\cos(\omega_0\, t) - \rho}{\sin(\omega_0\, t)}\right)\right]. \tag{13}$$

Furthermore, the *relative radial acceleration* is

$$\begin{aligned}
a_R(t) &= \frac{\mathrm{d}}{\mathrm{d}t}\, v\,\cos\theta_R(t) \\
&= -v\,\dot\theta_R(t)\,\sin\theta_R(t) \\
&= -v\,\omega_0\,\rho\,\frac{\rho - \cos(\omega_0\, t)}{1 - 2\rho\cos(\omega_0 t) + \rho^2}\,\sin\left[\omega_0\, t + \tan^{-1}\left(\frac{\cos(\omega_0\, t) - \rho}{\sin(\omega_0\, t)}\right)\right]
\end{aligned} \tag{14}$$

In Fig. 3, the case of a descending satellite is considered. In Fig. 3a, angles $E(t)$, $\phi(t)$ and $\theta_R(t)$, as functions of time measured in hours are reported. The considered time interval is such that the elevation angle $E(t)$ ranges from 90.00 to 5.1603

Fig. 3 Descending satellite: **a** Elevation angle $E(t)$, earth central angle $\phi(t)$, and radial angle $\theta_R(t)$, in degrees, as functions of time; **b** distance $R(t)$ as function of time; **c** radial speed $v_R(t)$ as functions of time

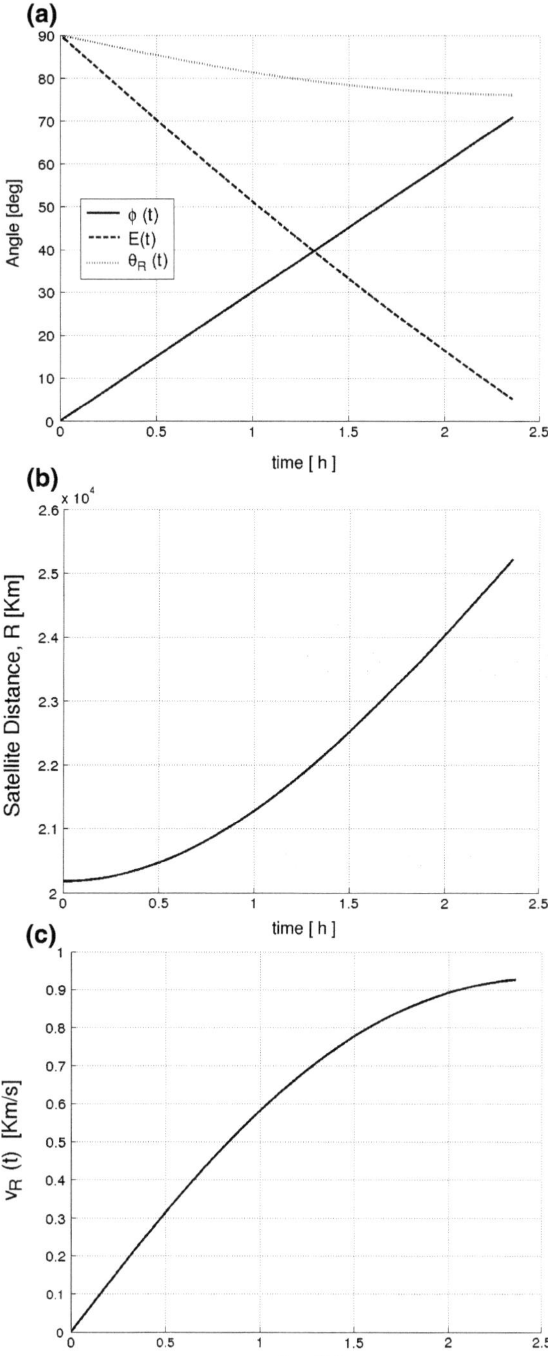

degrees which is the minimum elevation angle such that the satellite can be considered observable. In Fig. 3b, $R(t)$ is reported as function of time. According to these behaviors and expression (13), in Fig. 3c, the relative radial speed $v_R(t)$ is reported as function of time.

The length of the observation intervals of interest in the applications varies form 1 ms for the first coarse acquisition (Tsui 2000, Chap. 6, pp. 135–149) to hundreds of milliseconds for indoor applications (Seco-Granados et al. 2012). From Fig. 3c it is evident that within these observation intervals the relative radial speed can be considered constant. Consequently, the relative radial acceleration is negligible. Analogous results are obtained for an ascending satellite.

3 Propagation Channel

The propagation channel will be derived under the following assumptions.

1. Moving TX and stationary RX.
2. Free-space propagation and far-field condition (Napolitano 2012, Sect. 7.1.1).
3. Wide-band transmit and receive antennas (Napolitano 2012, pp. 383–386).
4. Constant relative radial speed within the observation interval (see Sect. 2).
5. Relativistic time dilation between clocks in relative motion due to Lorentz transformations neglected. Relativistic effect due to the different gravitational potential between TX and RX neglected. [In order to compensate both these effects, in the GPS system the satellite clock frequency in adjusted to 10.22999999543 MHz prior to launch (Kaplan and Hegarthy 2006, Sect. 7.2.3, pp. 306–308). Consequently, the frequency observed by the user at sea level is 10.23 MHz.]
6. Relativistic correction due to the so–called Sagnac effect, also known as Earth rotational effect, neglected assuming an inertial reference frame (Kaplan and Hegarthy 2006, Sect. 7.2.3, pp. 306–308).
7. Atmospheric perturbations due to the presence of the troposphere and ionosphere not taken into account. [In the GPS system, in the case of single–frequency receiver, models of the ionosphere are employed to correct the ionospheric delay and scintillation on the pseudo–range measurements (GPS Directorate 2013). The correction parameters are broadcast with the Navigation Message and periodically updated by the monitor stations of the control segment.]

Under assumptions 1–3, for a transmitted signal $z_{TX}(t)$, the received signal is

$$z_{RX}(t) = A(t)\, z_{TX}(t - D(t)) \tag{15}$$

where $D(t)$ is the time-varying delay which is linked to the time-varying distance $R(t) = \|S(t) - P\|$ by $c\,D(t) = R(t - D(t))$ (Napolitano 2012, Sect. 7.1.5) and $A(t)$ is a time-varying amplitude that can be considered constant under mild conditions (Napolitano 2012, pp. 387, 396, 403).

Let

$$z_{\text{TX}}(t) = \text{Re}\left[x(t)\,e^{j2\pi f_c t}\right] \tag{16}$$

be the transmitted signal with carrier frequency f_c. Under the assumption of constant relative radial speed within the observation interval (assumption 4) the time-varying delay is a linear function of time. Consequently we have

$$z_{\text{RX}}(t) = \text{Re}\left[y(t)\,e^{j2\pi f_c t}\right] \tag{17}$$

with

$$y(t) = b\,x(st - d)\,e^{j2\pi vt} \tag{18}$$

where b is the complex gain, s the time-scale factor, v the frequency shift, and d the delay. For a moving TX and stationary RX, $s = c/(c + v_R)$, $v = (s - 1)f_c$, $d = R(0)/(c + v_R)$ (Napolitano 2012, Sect. 7.3.3).

From (18) it follows that the Doppler effect produces a time stretch of the transmitted signal. If the so-called narrow-band condition is satisfied, that is, if

$$B\,T \ll 1/|1 - s| \simeq c/|v_R| \tag{19}$$

where B is the signal bandwidth, T the observation interval, and $c \simeq 3 \times 10^8 \text{m s}^{-1}$ the medium propagation speed, then the time-scale factor s can be considered unit in the argument of $x(\cdot)$ (but not in the complex exponential) and the Doppler effect reduces just to a frequency shift of the carrier, which is a well known and widely adopted model (Napolitano 2012, Sect. 7.5; Van Trees 1971, pp. 240–242).

If the simplifying assumptions 5–7 are not satisfied, the model for the received signal is still (18). In fact, the relativistic time dilation is an affine transformation and all other corrections to the satellite motion do not modify the constant characteristic of the radial speed within observation intervals of interest in the applications, that is, up to hundreds of milliseconds. Of course, the values of s, v, d, and b are different with respect to those for the ideal case of moving TX with circular orbit and stationary RX. Therefore, in *estimation procedures* based on model (18), also the effects of lack of validity of the simplifying assumptions 5–7 are accounted for. Moreover, also the case of moving RX can be described by model (18) (Napolitano 2012, Sect. 7.3.1).

4 Transmitted Signal

4.1 Signal Model

The continuous-time GPS-L1 signal (GPS Wing 2010, Sects. 3.2–3.3, pp. 3–17), (Kaplan and Hegarthy 2006, Chap. 4, pp. 113–116) is a quadrature phase-shift keying (QPSK) signal

$$z_{\mathrm{TX}}(t) = \sqrt{2}\, A\, d(t)\, c(t)\, \cos(2\pi f_{\mathrm{L1}} t + \phi_0)$$
$$+ A\, d(t)\, p(t)\, \sin(2\pi f_{\mathrm{L1}} t + \phi_0) \qquad (20)$$

where $f_{\mathrm{L1}} = 1575.42\,\mathrm{MHz}$ is the $L1$ carrier frequency. In (20):

(1) $d(t)$ is the *navigation message*. It is obtained by interleaving two periodic components with periods T_{frame} (frame period) and $T_{\mathrm{page}} = 5 T_{\mathrm{frame}}$ (page period), respectively, two binary pulse-amplitude-modulated (PAM) signals with bit period T_{b} such that $T_{\mathrm{frame}} = 300 T_{\mathrm{b}}$; the two PAM signals are multiplied by periodic signals with periods T_{page} and T_{frame} (GPS Wing 2010, Sect. 20.3, pp. 68–122).

(2) $c(t)$ is the *coarse acquisition (C/A) code* signal which is the ranging signal for civil applications. It is a relatively short code to enable a rapid acquisition. The signal $c(t)$ is the periodic replication with period $T_{\mathrm{CA}} = N_{\mathrm{c}} T_{\mathrm{c}}$ of a fixed Gold sequence with chip period T_c and $N_{\mathrm{c}} = 1023$ chips that identifies the satellite (GPS Wing 2010, Sect. 3.3, pp. 26–30). It results $T_c = 0.9775\,\mu\mathrm{s}$, $T_{\mathrm{CA}} = 1\,\mathrm{ms}$, and $T_{\mathrm{b}} = 20 N_{\mathrm{c}} T_{\mathrm{c}}$.

(3) $p(t)$ is the *precision P(Y) code* signal. It is a periodic signal with period equal to 1 week obtained by periodic replication of a fixed pseudo-noise (PN) sequence (GPS Wing 2010, Sect. 3.3, pp. 18–25). It is modeled as a binary PAM signal with i.i.d. symbols and bit period $T_{\mathrm{p}} = T_{\mathrm{c}}/10$ within realistic observation intervals.

4.2 Second-Order Cyclostationarity

Due to the presence in (20) of periodic replication operations and PAM signals, in (Napolitano and Perna 2013a), it is shown that the complex envelope signal associated to $z_{\mathrm{TX}}(t)$

$$x(t) = \sqrt{2}\, A\, d(t)\, c(t) - j\, A\, d(t)\, p(t) \qquad (21)$$

is second-order wide-sense cyclostationary. That is, its expected value and autocorrelation function are periodic functions of time. The analytical expression of (conjugate) cyclic autocorrelation functions and (conjugate) cyclic spectra of $x(t)$ are derived in (Napolitano and Perna 2013a). These expressions are very complicate due to the complex structure of the GPS-L1 signal (20). In (Napolitano and Perna 2013a), it is shown that different signal models should be considered depending on the length T of the observation interval. Consequently, different cyclic features are evidenced depending on the value of T.

(1) Let $T = T_{\mathrm{CA}} = 1\,\mathrm{ms}$, that is, the length of the observation interval is coincident with the width T_{CA} of the coarse acquisition interval. This is the minimum observation interval during the first step of the L1 signal acquisition to synchronize the C/A code with a replica locally generated by the receiver and aimed at identifying the satellite and estimating the Doppler shift (Borre et al. 2007,

Chap. 5, pp. 69–72; Pi and Huang 2010; Tsui 2000, Chap. 7, pp. 135–149). In this case, assuming that $d(t) = 1$ and no data-bit transition is present within the observation interval, the signal model for the GPS-L1 signal is

$$x(t) = \sqrt{2}A\, c(t) - jA\, p(t) \tag{22}$$

where the C/A code signal $c(t)$ must be modeled as a binary PAM signal with bit period T_c since no replication (with period T_{CA}) of this signal can be observed within an observation interval $T = T_{\mathrm{CA}}$. Thus, this signal model for $c(t)$ exhibits cyclostationarity at cycle frequencies $\alpha = k/T_c$, k integer. The signal $p(t)$ is a PAM signal with period $T_p = T_c/10$. It has cycle frequencies $\alpha = k/T_p$, k integer, which are a subset of the cycle frequencies $\alpha = k/T_c$. The cycle frequencies $\alpha = k/T_c$ are pure second-order cycle frequencies (Gardner and Spooner 1994). Removing the periodic component with period T_{CA} has no sense since the observation interval is coincident with this period. Therefore, the signal $x(t)$ should be modeled as cyclostationary with period T_c.

(2) Let be $T = 20\,T_{\mathrm{CA}} = T_b = 20$ ms. This is the observation interval for the second step of the L1 signal acquisition, the tracking phase to synchronize the navigation bit (Borre et al. 2007, Chap. 5, pp. 72–73; Tsui 2000, Chap. 9, pp. 193–198), or the observation interval for dual frequency GPS L1/L2C receivers (Qaisar and Dempster 2011). If the observation interval is such that no data-bit transition is present, then the model for the GPS-L1 signal is that in (22), where, unlike the case $T = T_{\mathrm{CA}}$, the signal $c(t)$ must be modeled as the periodic replication with period T_{CA} of a deterministic signal $u_c(t)$. The signal exhibits cyclostationarity with cycle frequencies $\alpha = k/T_c = kN_c/T_{\mathrm{CA}}$ which, in such a case, are impure second-order cycle frequencies (Gardner and Spooner 1994) and disappear if the additive periodic component (the expected value) is removed from $x(t)$ before computing the cyclic statistics. Therefore, the signal $x(t)$ should be modeled as cyclostationary with period T_c while the signal $x(t)$ with the periodic component removed is cyclostationary with period T_p.

(3) Let be $T = 400\,T_{\mathrm{CA}} = 20\,T_b = 400$ ms. This value of coherent integration time can be adopted to improve performance in indoor environment (Seco-Granados et al. 2012). Since the modulation of $d(t)$ must be considered in both in-phase and in-quadrature components of $x(t)$, in correspondence of this data-record length, the model for the GPS-L1 signal does not contain a significant additive periodic component unlike the case $T = 20\,T_{\mathrm{CA}} = T_b = 20$ ms. Therefore, both $x(t)$ and $x(t)$ with the periodic component removed are cyclostationary with period T_c.

In all considered cases, the periods of periodic functions or PAM signals are multiples of T_p, the smallest period equal to the reciprocal of the clock frequency. Thus, in all cases the signal $x(t)$ is cyclostationary. Let us denote by T_{cs} the period of cyclostationarity. Accordingly with the previous discussion, such a value depends on the signal model which, in turn, depends on the observation interval T. It results $T_{\mathrm{cs}} = T_c$ if the additive periodic component is not removed from $x(t)$.

If such a periodic component is removed from $x(t)$, then $T_{cs} = T_c$ for $T = 1$ ms and $T = 400$ ms and $T_{cs} = T_p$ for $T = 20$ ms.

Both autocorrelation function and conjugate autocorrelation function are necessary for a complete second-order characterization in the wide sense of complex-valued signals (Schreier and Scharf 2003). In the following, notation $(*)$ will be adopted for an optional complex conjugation in order to consider, in the same formula, both cyclic statistics and conjugate cyclic statistics. In addition, $(-)$ will be an optional minus sign linked to $(*)$.

For the cyclostationary process $x(t)$ the *(conjugate) autocorrelation function* (Gardner et al. 2006) is given by

$$\mathrm{E}\left\{x(t+\tau)\,x^{(*)}(t)\right\} = \sum_{k=-\infty}^{+\infty} R_{xx^{(*)}}^{k/T_{cs}}(\tau)\,e^{j2\pi(k/T_{cs})t} \tag{23}$$

with

$$R_{xx^{(*)}}^{\alpha}(\tau) \triangleq \lim_{T\to\infty}\frac{1}{T}\int_{-T/2}^{T/2} \mathrm{E}\left\{x(t+\tau)\,x^{(*)}(t)\right\}\,e^{-j2\pi\alpha t}\mathrm{d}t \qquad \alpha = \frac{k}{T_{cs}},\ k\in\mathbb{Z} \tag{24}$$

referred to as *(conjugate) cyclic autocorrelation functions*. The *Loève bifrequency spectrum* is given by

$$\mathrm{E}\left\{X(f_1)\,X^{(*)}(f_2)\right\} = \sum_{k=-\infty}^{+\infty} S_{xx^{(*)}}^{k/T_{cs}}(f_1)\,\delta\left(f_2 + (-)(f_1 - k/T_{cs})\right) \tag{25}$$

with

$$S_{xx^{(*)}}^{\alpha}(f) \triangleq \int_{\mathbb{R}} R_{xx^{(*)}}^{\alpha}(\tau)\,e^{-j2\pi f\tau}\mathrm{d}\tau \qquad \alpha = \frac{k}{T_{cs}},\ k\in\mathbb{Z} \tag{26}$$

called *(conjugate) cyclic spectra*. In (25), $\delta(\cdot)$ is Dirac delta and $X(f)$ denotes the Fourier transform defined in a distributional sense (Gel'fand and Vilenkin 1964, Chap. 6; Napolitano 2012, Sects. 1.1.2, 4.2.1).

5 Received Signal

Assuming for the bandpass GPS-L1 signal an approximate bandwidth $B \simeq 1/T_p = 10.23$ MHz, where T_p is the width of the narrowest rectangular pulse in the signal model, we have that $T = 1$ ms $\Rightarrow BT \simeq 10.23 \times 10^3$ and $T = 10$ ms $\Rightarrow BT \simeq 1.02 \times 10^5$. By considering $|v_R| = 0.9$ kms^{-1} (see Fig. 3c) it results $c/|v_R| \simeq 3.3 \times 10^5$ and $|1-s| \simeq 3 \times 10^{-6}$. Thus, the narrow-band condition (19) is practically satisfied for $T = 1$ ms and is not satisfied for $T \geq 10$ ms.

From (18), it follows that if $x(t)$ exhibits cyclostationarity with cycle frequency α, then $y(t)$ exhibits cyclostationarity with cycle frequency $s\alpha$. The cyclic autocorrelation functions and cyclic spectra of $x(t)$ and $y(t)$ are linked by

$$R_{yy^*}^{s\alpha}(\tau) = |b|^2 e^{j2\pi v\tau}\, e^{-j2\pi\alpha d}\, R_{xx^*}^{\alpha}(s\tau) \tag{27}$$

$$S_{yy^*}^{s\alpha}(f) = |b|^2 e^{-j2\pi\alpha d}\, \frac{1}{|s|} S_{xx^*}^{\alpha}\left(\frac{f-v}{s}\right). \tag{28}$$

respectively. Moreover, if $x(t)$ exhibits conjugate cyclostationarity with conjugate cycle frequency β, then $y(t)$ exhibits conjugate cyclostationarity with conjugate cycle frequency $s\beta + 2v$ and the conjugate cyclic autocorrelation functions and conjugate cyclic spectra of $y(t)$ and $x(t)$ are linked by

$$R_{yy}^{s\beta+2v}(\tau) = b^2\, e^{j2\pi v\tau}\, e^{-j2\pi\beta d}\, R_{xx}^{\beta}(s\tau) \tag{29}$$

$$S_{yy}^{s\beta+2v}(f) = b^2\, e^{-j2\pi\beta d}\, \frac{1}{|s|} S_{xx}^{\beta}\left(\frac{f-v}{s}\right). \tag{30}$$

respectively.

From (27–30) it follows that results of Sect. 4.2 for the complex signal $x(t)$ can be applied to the received signal $y(t)$ with the obvious modifications of the (conjugate) cycle frequencies. Specifically, the received signal exhibits cyclostationarity with cycle frequencies sk/T_{cs}, $k \in \mathbb{Z}$ and conjugate cyclostationarity with conjugate cycle frequencies $sk/T_{cs} + 2v$, $k \in \mathbb{Z}$.

Due to the presence of the non unit time-scale factor s in the expression (18) of the received signal $y(t)$ in terms of $x(t)$, it follows that $y(t)$ and $x(t)$ are not jointly cyclostationary but, rather, jointly spectrally correlated (Napolitano 2012, Chap. 4). In fact, accounting for the Fourier transform (defined in a distributional sense) of both sides of (18)

$$Y(f) = \frac{b}{|s|}\, X\left(\frac{f-v}{s}\right) e^{-j2\pi(f-v)d/s} \tag{31}$$

one obtains the Loève bifrequency cross-spectrum of $y(t)$ and $x(t)$

$$\mathrm{E}\left\{Y(f_1)\, X^{(*)}(f_2)\right\} = \frac{b}{|s|}\, e^{-j2\pi(f_1-v)d/s}$$

$$\sum_{k=-\infty}^{+\infty} S_{xx^{(*)}}^{k/T_{cs}}\left(\frac{f_1-v}{s}\right) \delta\left(f_2 - (-)\left(\frac{k}{T_{cs}} - \frac{f_1-v}{s}\right)\right) \tag{32}$$

It is constituted by spectral masses concentrated on a countable set of lines with non unit slope ($s \neq 1$).

Starting from (32) one obtains the cross-correlation function of $y(t)$ and $x(t)$

$$\mathrm{E}\left\{y(t+\tau)\,x^{(*)}(t)\right\}$$

$$= b\,e^{j2\pi v\tau}\sum_{k=-\infty}^{+\infty} R_{xx^{(*)}}^{k/T_{\mathrm{cs}}}\!\left((s-1)t+s\tau-d\right)e^{j2\pi(k/T_{\mathrm{cs}}+v)t}\,. \tag{33}$$

Due to the presence of the nonunit time-scale factor s in the argument of $R_{xx^{(*)}}^{k/T_{\mathrm{cs}}}(\cdot)$, the (conjugate) cross-correlation function does not contain any finite-strength additive sinewave component provided that the functions $R_{xx^{(*)}}^{k/T_{\mathrm{cs}}}(\cdot)$ are summable, that is, in all those cases where k/T_{cs} are pure cycle frequencies (see Sect. 4.2). In contrast, if $x(t)$ contains an additive periodic component, then (33) is the expression of the (conjugate) cross-correlation function of $y(t)$ and $x(t)$ with its periodic component removed. In such a case, from (33) it follows that, accordingly with (32), the processes $y(t)$ and $x(t)$ are not jointly (almost-)cyclostationary.

It is worthwhile to underline that the analytical model of the received signal (18) is not influenced by the simplifying assumptions made on the satellite orbit in Sect. 2. Consequently, the expressions (27–30) for the cyclic statistics of the received signal and of the cross statistical characterization of transmitted and received signals (32, 33) are not approximate relationships. In contrast, the values of the parameters b, s, v, and d depend of the effective orbit of the satellite.

In (Napolitano and Perna 2013b), a synchronization technique for the coarse acquisition code is proposed that models the transmitted and received signals as jointly spectrally correlated. Performance is evaluated in terms of sample root mean squared error (rmse) of the estimates of parameters s, v, d, and $\phi = \angle b$. The proposed technique is shown to perform slightly worst than the classical synchronization method based on the narrow-band cross-ambiguity function (NB-CAF) when the data-record length is 1 ms (corresponding to $N_b = 1{,}023$ bits of the coarse acquisition code), which is the value adopted in commercial GPS receivers and such that the narrow-band condition (19) is satisfied. In contrast, when observation intervals larger than 1 ms are adopted aimed at obtaining a beneficial effect toward noise and interference, the proposed method provides a significant performance improvement. A scenario is considered with additive white Gaussian noise (AWGN) with signal-to-noise ratio (SNR) equal to 0 dB in the bandwidth $(-f_s/2, f_s/2)$, where f_s is the sampling frequency, and with an interfering binary phase-shift-keying (BPSK) signal with carrier frequency $f_{\mathrm{L}1}$ and signal-to-interference ratio (SIR) equal to 5 dB. For $N_b = 8{,}184$ bits, the rmse of the estimates of v, d, and ϕ by the technique proposed in (Napolitano and Perna 2013b) can be up to two orders of magnitude smaller than those obtained by the NB-CAF at $N_b = 1{,}023$ and $N_b = 8{,}184$. The gain in performance is obtained since for $N_b = 8{,}184$ the narrow-band condition (19) is not satisfied and, hence, the NB-CAF method is based on the wrong model that assumes $s = 1$ in the argument of the complex envelope $x(\cdot)$ in (18). In contrast, the technique proposed in (Napolitano and Perna 2013b) benefits of the increased data-record length and the consistency of the estimators of the involved statistical functions.

6 Conclusion

The effects of the satellite motion on the received signal on the Earth are analyzed for GPS. In time intervals of interest in the applications, the relative radial speed between satellite and ground receiver can be considered constant. Under these conditions, the transmitted cyclostationary signal is still cyclostationary at receiver, but with different cycle frequencies and cyclic features. Moreover, the transmitted and received signals are jointly spectrally correlated. They can be modeled as jointly cyclostationary only if the data-record length does not exceeds 1 ms. However, such a data-record length is not sufficient in indoor applications and in the presence of strong jammer. Significantly longer data-record lengths can be used by the adoption of the spectrally correlated model allowing to get significantly better performance in the presence of severe noise and interference environments.

References

Borre K, Akos DM, Bertelsen N, Rinder P, Jensen SH (2007) A software-defined GPS and galileo receiver. Birkhauser, Boston

Gardner WA, Spooner CM (1994) The cumulant theory of cyclostationary time-series, part I: foundation. IEEE Trans Signal Process 42:3387–3408

Gardner WA, Napolitano A, Paura L (2006) Cyclostationarity: half a century of research. Signal Process 86(4):639–697

Gel'fand IM, Vilenkin NY (1964) Generalized functions, applications of harmonic analysis, vol 4. Academic Press, New York

GPS Directorate (2013) Interface Specification IS GPS 200, Navstar GPS Space Segment/Navigation User Interfaces (IS GPS 200G). Global Positioning Systems Directorate, Systems Engineering and Integration

GPS Wing (2010) Navstar GPS Space Segment/Navigation User Interfaces (IS GPS 200E). GPS Wing (GPSW), Space & Missiles Center (SMC)—GPSW LAAFB

Kaplan E, Hegarthy CJ (2006) Understanding GPS—principles and applications, 2nd edn. Artech House, Boston

Maral G, Bousquet M (2009) Satellite communications systems—systems, techniques and technologies, 5th edn. Wiley, London

Napolitano A (2012) Generalizations of cyclostationary signal processing: spectral analysis and applications. Wiley–IEEE Press, Chichester

Napolitano A, Perna I (2013a) Cyclic spectral analysis of the GPS signal. submitted to Digital Signal Processing

Napolitano A, Perna I (2013b) On synchronization of Doppler-stretched GPS signals. In: 21st European signal processing conference (EUSIPCO 2013). Marrakech, Morocco

Pi Y, Huang P (2010) Research on novel structure of GPS signal acquisition based on software receiver. In: International symposium on intelligent signal processing and communication systems (ISPACS)

Qaisar SU, Dempster AG (2011) Cross-correlation performance assessment of global positioning system (GPS) L1 and L2 civil codes for signal acquisition. IET Radar Sonar Navig 5(3):195–203

Schreier PJ, Scharf LL (2003) Second-order analysis of improper complex random vectors and processes. IEEE Trans Signal Process 51(3):714–725

Seco-Granados G, Lopez-Salcedo J, Jimenez-Banos D, Lopez-Risueno G (2012) Challenges in indoor global navigation satellite systems: Unveiling its core features in signal processing. IEEE Signal Process Mag 29(2):108–131
Tsui JBY (2000) Fundamentals of global positioning system receivers. Wiley, New York
Van Trees HL (1971) Detection, estimation, and modulation theory, Part III. Wiley, New York

Cyclostationary Processing of Vibration and Acoustic Emissions for Machine Failure Diagnosis

Cristián Molina Vicuña and David Quezada Acuña

Abstract The use of vibrations and other variables to infere the mechanical condition of machines is a common practice nowadays. Several equipments for measurement and processing are available. Most of them offer FFT spectrum as the main tool for analysis, thus allowing the assessment of the stationary part of the signal only. Although this might be sufficient in some cases, it is certainly inappropriated in others. Recent advances in signal processing have opened the possibilities for analyzing a special type of non-stationary signals, called cyclostationary signals. It has also been shown that the behaviour of machines can be highly cyclostationary in some cases. Moreover, being stationarity a special case of cyclostarionarity, the advantages of the cyclostationary approach become evident. Still, even under consideration of these facts, the use of cyclostationarity appears to be still restricted to the scientific community, being its use in the industry far from being a reality. This chapter presents the concept of cyclostationarity, its terminology and its relation with traditional signal processing tools in a descriptive way. Two examples of real data analysis from the cyclostationary viewpoint are also presented.

1 Introduction

Machine Condition Monitoring (CM) has been used for decades as part of maintenance strategies in different industries. Nowadays, the use of vibration-based CM systems (off-line and on-line) is wide spread on a large variety of machines. In the majority of cases, the diagnosis of the machine's health relies on the assessment of

C. M. Vicuña (✉) · D. Q. Acuña
Laboratorio de Vibraciones Mecánicas, Universidad de Concepción, Edmundo Larenas 270 (int.), Concepción, Chile
e-mail: crimolin@udec.cl

D. Q. Acuña
e-mail: davque@gmail.com

F. Chaari et al. (eds.), *Cyclostationarity: Theory and Methods*,
Lecture Notes in Mechanical Engineering, DOI: 10.1007/978-3-319-04187-2_10,
© Springer International Publishing Switzerland 2014

the Fourier magnitude spectrum of the measured signals, meaning they are inherently assumed to be stationary. Considering the large quantity of physical processes occurring inside a machine and all parameters affecting the resulting vibrations, it is easy to see that strict stationarity is a *hard-to-meet* property for the signals. In some cases the vibration produced is, however, *mainly* stationary (e.g. the vibration resulting from imbalance and eccentricity). In other cases, the resulting vibrations are considerably non-stationary (e.g. vibrations from compressors, IC engines, and from defective bearings). There are also phenomena producing vibrations with one portion being mainly stationary and other, non-stationary. This is, for example, the case of gear vibrations, where the low-frequency region is dominated by the stationary part; whereas the high-frequency region is mostly non-stationary (Antoni and Bonnardot et al. 2004; Capdessus et al. 2000; Raad and Antoni 2008). In fact, high-frequency phenomena which repeats in an apparently periodic fashion may very unlikely be stationary, because their dependence on the interaction of elements at a microscopic level produce random variations from cycle to cycle. This particularly applies to Acoustic Emissions (AE), high-frequency structure-borne waves sometimes used for CM.

Cyclostationarity (CS) is a property of a group of signals whose statistical properties vary periodically with time—being therefore non-stationary—, including stationarity as a particular case. CS is of special interest in machine CM, because vibrations resulting from different machines and processes have CS properties. These include cases in which periodic modulations of random vibrations occur, or where impact series exist, as in IC engines (Antoni et al. 2002), high frequency vibrations from gears (Antoni and Bonnardot et al. 2004), forging machines, etc. and virtually all AE containing bursts (Vicuna and Hoeweler 2013). The most symbolic case found in the literature refers to bearings with localized defects, a situation where a succession of impacts excite the impulse response of the system in an apparently periodic manner, which is actually random due to slip ocurring between the bearing components (Ho and Randall 2000; Randall 2001; Antoni and Randall 2002).

Despite the rich theoretical background and the promising results that can be obtained from the CS treatment of vibrations and AE, there is still low usage of CS signal processing tools for vibration analysis. We present in this work some applications of CS analysis of machine vibrations and AE, aiming to illustrate some of the benefits of the CS treatment and assessment of mechanical signals. A brief overview of the CS theory and terminology is also presented, which is necessary to understand the discussions of the applications.

2 Cyclostationarity

This section provides an overview of the CS theory, terminology and signal processing tools.

2.1 Definition and Terminology

Cyclostationary signals are a type of non-stationary signals whose statistical properties change periodically with time (or angle of rotation). This is in opposition to stationary signals, whose statistical properties remain constant with time; and to random signals, whose statistical properties change (but not periodically) with time. Cyclostationary signals are generally random in their waveform, but exhibit some hidden periodicity in its energy flow, which is generated by some periodic process, whose period is called *cycle*. Its inverse is called *cyclic frequency* and is denoted by the symbol α. Formally, a signal *exhibits* cyclostationarity if exists a combination of linear and non-linear transformations that produces periodic components as a result. It is said that a signal exhibits cyclostationarity at the cyclic frequency α, if there is a combination of linear and non-linear transformations that produce a pure sinusoidal with frequency α.

Based on the definition above, the order of cyclostationarity is determined by the maximum order of the transformations used to obtain periodic components. According to this (Antoni and Bonnardot et al. 2004):

1. A signal is *first-order cyclostationary (CS1)* with fundamental cycle T, if only a linear transformation is sufficient to obtain a periodic signal. In particular, if the first-order moment is periodic with fundamental period T:

$$m_X(t) \triangleq \mathbb{E}\{X(t)\} = m_X(t+T) \tag{1}$$

   CS1 signals show periodic time histories, in which additive random stationary background noise can be present.
2. A signal is *second-order cyclostationary (CS2)* with fundamental cycle T, if the order of the non-linear transformation needed to obtain a periodic signal is two. In particular, if its second-order moment (i.e. the autocorrelation function, ACF) is periodic with fundamental period T:

$$R_{XX}(t_1, t_2) \triangleq \mathbb{E}\{X(t_1) X^*(t_2)\} = R_{XX}(t_1 + T, t_2 + T) \tag{2}$$

   CS2 signals are stochastic signals undergoing periodic modulations.
3. A signal is *n-th-order cyclostationary (CSn)* with fundamental cycle T, if the order of the non-linear transformation needed to obtain a periodic signal is $n > 2$. Such signal is also called high-order cyclostationary (Spooner 1994).

A process which is both CS1 and CS2 is called *wide-sense* cyclostationary; whereas a process whose moments till infinity are periodic is called *strict-sense* cyclostationary.

The definitions 1, 2 and 3 assume the existence of a single periodic process with fundamental cycle T. However, a signal can contain several hidden periodic processes with different fundamental cycles. Accordingly, it is distinguished between a signal which *exhibits* cyclostationarity and a signal which *is* cyclostationary at a given order.

Denoting by $\mathscr{A}$ the set of all cyclic frequencies α present in the signal obtained after the appropriate linear and non-linear transformations:

- A signal *exhibits* cyclostationarity at a cyclic frequency α, if the set $\mathscr{A}$ contains α and its multiples, *among other* cyclic frequencies.
- A signal *is* cyclostationary at a cyclic frequency α, if the set $\mathscr{A}$ contains *only* α and its multiples.

Moreover, if for a signal which exhibits cyclostationarity, the cyclic frequencies contained in $\mathscr{A}$ are such that they share no integer common factor, the signal is said to be *poly*-cyclostationary.

Also from the previous definitions 1, 2 and 3, it is noted that a signal which is first-order cyclostationary will be cyclostationary at all orders. For example, the ACF of a periodic signal $p(t)$ is:

$$R_{PP}(t_1, t_2) \triangleq \mathbb{E}\left\{p(t_1)\, p^*(t_2)\right\} = p(t_1)\, p^*(t_2)$$
$$= m_P(t_1)\, m_P^*(t_2) \tag{3}$$

The ACF in Eq. 3 is periodic, because it depends on the first-order moment $m_p(t)$, which is periodic. The same occurs for higher orders. Hence, all moments of a periodic signal are periodic. However, the periodicity of the moments of order $n > 1$ is only a result of the periodicity of the first-order moment. Accordingly, a periodic signal is *pure* cyclostationary at the first order (denoted PCS1), but *impure* at higher orders. Distinguishing between pure and impure cyclostationarity is important, because a PCS1 signal does not require the cyclostationary approach for its analysis, since the classical stationary approach suffices.

To discriminate between pure and impure cyclostationarity, the cumulant functions—instead of the moments—are used. For example, the cumulant function of second order (i.e. the autocovariance function, ACVF) substracts the impure terms induced by the first order moment, thus allowing to evaluate the pure second-order cyclostationary content of the signal. The same holds for higher orders, where higher-order cumulant functions are used (Spooner 1994; Gardner 1994; Spooner and Gardner 1994). In the example of the periodic signal, its ACVF reads:

$$K_{PP}(t_1, t_2) \triangleq \mathbb{E}\left\{[p(t_1) - m_P(t_1)]\left[p^*(t_2) - m_P^*(t_2)\right]\right\}$$
$$= R_{PP}(t_1, t_2) - m_P(t_1)\, m_P^*(t_2) = 0 \tag{4}$$

This result reflects no cyclostationarity at the second order for the periodic signal $p(t)$. Accordingly, the ACVF should be preferred instead of the ACF to evaluate if a process is PCS2.

2.2 Why Cyclostationarity?

The cyclostationary approach presents several advantages for the fault diagnosis of mechanical systems through the assessment of variables such as vibrations and AE. This comes from the fundamental fact that the observed signals hardly meet the stationary condition, which is inherently assumed when they are processed with conventional tools such as the Fourier transform or the power spectral density (PSD). Cyclostationarity encompasses a larger family of signals and, therefore, is a more general and powerful approach than the stationary approach.

Important results have been obtained in the application of cyclostationarity to machine CM. For example, in the field of gear diagnosis, it has been found that some failures like spalling produce periodic modulations in the signal which are well detected using the cyclostationary approach (Capdessus et al. 2000). Interestingly, it has been suggested that low-frequency vibrations measured in gears behave as CS1 signals, whereas high-frequency vibrations present PCS2 characteristics. Moreover, it has been argued that the former is a result of the macro phenomena, and the latter is a result of the micro-phenomena involved in the gearing process (Antoni and Bonnardot et al. 2004; Raad and Antoni 2008). This is in perfect accordance with results from our researches involving AE measurements, as presented in Sect. 3.2. The field of bearing diagnosis is a classical example of vibration signals that are mostly stationary—mostly stationary noise—when the bearing suffers no failure, and are CS2 (or PolyCS2) under the presence of localized faults. A third example of cyclostationary signals from mechanical systems are the vibrations and sound observed in reciprocating machines (e.g. compressors, internal combustion engines). Such systems operate in a cyclic manner, undergoing a series of non-stationarities produced by different angle-locked events, such as openings and closings of valves.

2.3 Cyclostationary Signal Processing Tools

Cyclostationary signal processing (SP) tools exploit the special properties of cyclostationary signals to reveal their hidden periodicities, thus providing more information than the usual stationary SP tools. Again, if a signal is stationary, then no new information will be obtained by using cyclostationary SP tools.

The key idea of cyclostationary SP tools consists in decomposing the energy flow of filtered versions of the signal (covering the complete frequency range) not into constant values—as done in the PSD—, but into periodic values. The obtained time-frequency representation is called *instantaneous power spectrum*[1] and reveals the frequency structure of the energy flow in time. A natural step forward is the evaluation of the periodicities of the *instantaneous power spectrum*, which is accomplished by calculating its Fourier coefficients. Actually, it is this key step which reveals the hidden periodicities (if existent) of the signal. The collection of Fourier coef-

---

[1] Here we use the nomenclature presented in (Antoni 2009).

ficients is called *cyclic modulation spectrum*, and is a frequency-cyclic frequency representation.

Considering that the envelope of a signal is some function that envelopes the fluctuations of its energy flow as a function of time, the *instantaneous power spectrum* can be interpreted as a collection of envelopes of the band-pass-filtered versions of the signal. However, the introduction of the filterbank in the definition of the *instantaneous power spectrum* makes it a more powerful tool. Indeed, the integral of the *instantaneous power spectrum* along the frequency axis f merges the set of all envelopes into a single waveform similar to the classical envelope. The same relation is observed between the *cyclic modulation spectrum* and the envelope spectrum.

The *instantaneous power spectrum* and the *cyclic modulation spectrum* are affected by the uncertainty principle and, therefore, constitute no densities. As such, they do not conserve energy and are not unique. More advanced signal processing tools have been developed, which overcome this problem. Basically, they exploit the property that hidden periodicities produce correlation in the frequency domain. This property, also called *spectral redundancy* (Gardner 1991; Spooner 1994), provides the basis for the development of one of the most powerful cyclostationary SP tools: the *spectral correlation density (SCD)*. The SCD, mathematically defined as

$$
S_{xx}^{\alpha}\left(f\right) = \lim_{\Delta_f \to 0} \lim_{T \to \infty} \frac{1}{T\,\Delta_f} \int_T x_{\Delta_f}\left(t;\ f + \alpha/2\right) x_{\Delta_f}^{*}\left(t;\ f - \alpha/2\right) e^{-j2\pi\alpha t}\, dt
$$

(5)

is indeed a density, thus overcoming the problems of the *cyclic modulation spectrum*. It is recognized that as a particular case, when $\alpha = 0$ the SCD becomes the PSD. Energy normalisation of the SCD results in the *spectral coherence (SCoh)*, whose squared magnitude takes values only between 0 and 1 and provides a unitless measurement of the strength of correlation in the frequency domain. The SCoh is mathematically defined as

$$
\gamma_{xx}^{\alpha}\left(f\right) = \frac{S_{xx}^{\alpha}\left(f\right)}{\sqrt{S_{xx}^{0}\left(f + \alpha/2\right) S_{xx}^{0}\left(f - \alpha/2\right)}}
$$

(6)

Several estimators of the SCD and SCoh have been proposed, the most spread being the based in the *averaged cyclic periodogram* (Antoni 2007).

2.4 Relation with Conventional SP Tools

The relation between the *cyclic modulation spectrum* and the SCD is that the latter, being a density, is a fundamental constituent of the former (Antoni 2009). The same is valid between the *instantaneous power spectrum* and the quantity known as the *Wigner-Ville spectrum*. The Wigner-Ville spectrum is, in turn, related to the SCD in the same manner as the *instantaneous power spectrum* relates to the *cyclic modulation spectrum*. Additionally, the Wigner-Ville spectrum is related to the Wigner-Ville distribution, which is a (more) known time-frequency distribution, being equal to the mean value of it.

A formal proof of the relation between the SCD and the envelope spectrum is given in Randall (2001). It states that if the signal is dominated by the stochastic part (as is usually the case for the high frequencies, in particular for AE signals), the integral along f of the SCD is equivalent to the Fourier transform of the envelope of the squared signal, i.e.:

$$\int_{\mathbb{R}} S_{xx}^{\alpha}(f)\,df = \lim_{W \to \infty} \frac{1}{W} \int_{W} \mathbb{E}\left\{|x(t)|^2\right\} e^{-j2\pi\alpha t}\,dt \qquad (7)$$

This important result evidences that the well known envelope signal processing tool constitutes fundamentally a cyclostationary tool. Equation (7) provides a rich cyclostationary background to this classical tool. Furthermore, it shows that the envelope is a particular case of the SCD.

3 Applications of CS to Machine Diagnosis

This section shows two cases of real signals with CS features and how CS analysis can be used for extracting relevant information for machine condition monitoring.

3.1 Case 1: Faulty Bearing

This case deals with the vibration (acceleration) measured on a test bench with a bearing in healthy and faulty condition (light localized outer race defect). The outer race of the bearing is fixed and in both cases the inner race rotates at the same speed. The mean repetition rate of the impacts in the faulty case is BPFO = 94.8 Hz. Figure 1 shows the time history of the measured vibrations. At first sight, both signal look similar, being the increase of $\approx$33.7 % in the overall RMS in the faulty case, the most important difference. Some transients appear to be present, but is dificult to determine periodicities directly from the signal.

The presented signals are analyzed using the stationary approach. This is done by calculating the PSD spectrum of both signals (Fig. 2). The magnitude spectra (not shown) reveals the presence of discrete lines in the low frequency range. These components are caused by a periodic process in the machine, but are not related to the defect. There are no lines recognized at the fault frequency. Let us emphasize that the reason for this is that the vibration produced by the failure is not stationary, and therefore its presence is not revealed by the magnitude spectrum. Comparing both PSD spectra, a magnitude increase is observed in the faulty case. Note that the major

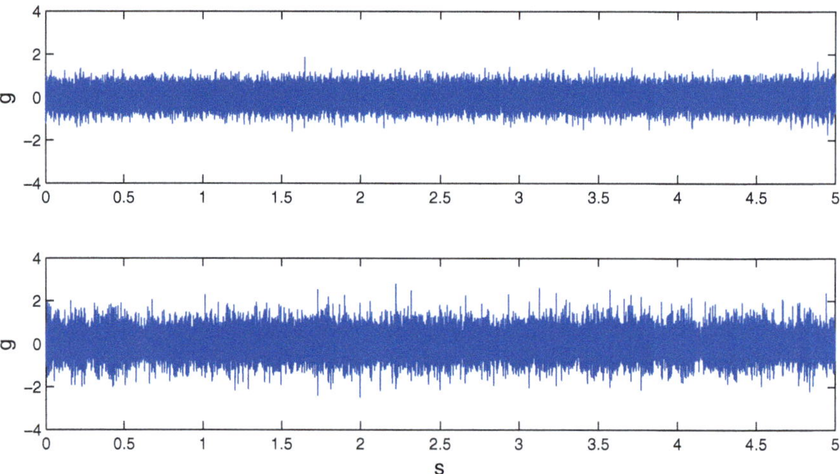

Fig. 1 Time history of the vibration measured without defect (*top*) and with defect (*bottom*)

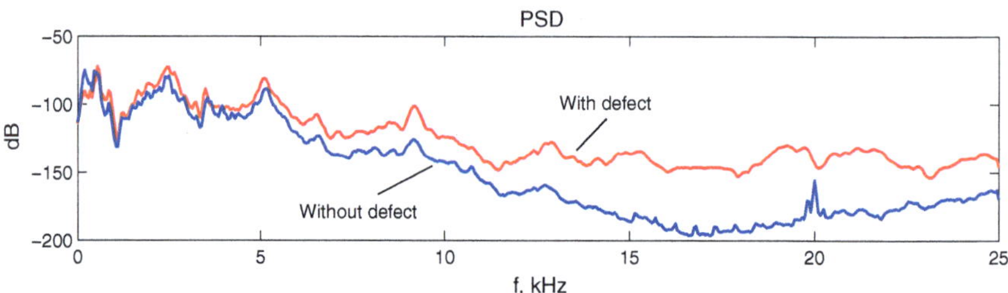

Fig. 2 Power spectral density of the faulty and non-faulty vibrations

change occurs in the high frequency region, above $\approx$6 kHz. This change could be attributed to the fault and, therefore, it is a common practice to use this information to filter the signal in this range previous to the calculation of the envelope spectrum. The result of this methodology is presented in Fig. 3, where the defect is revealed. Note we have purposely used the α symbol in the horizontal axis of the envelope spectrum to emphasize that the envelope is a CS quantity. It is actually this feature which permits to reveal the presence of the defect.

More information can be obtained by using more sophisticated CS tools. For example, Fig. 4 shows the SCoh calculated from both raw vibrations. It is observed that for the defect-free case, it presents non-zero components at some $\alpha \neq 0$ in a frequency range from 0 to $\approx$6 kHz and around $\approx$20 kHz. These components are due to the periodic part of the signal and are, therefore, impure second-order CS. They are present because in this case the auto-correlation function instead of the auto-covariance was used for the calculation of the SCoh. No other indications of second order CS are present in the signal for the defect-free case.

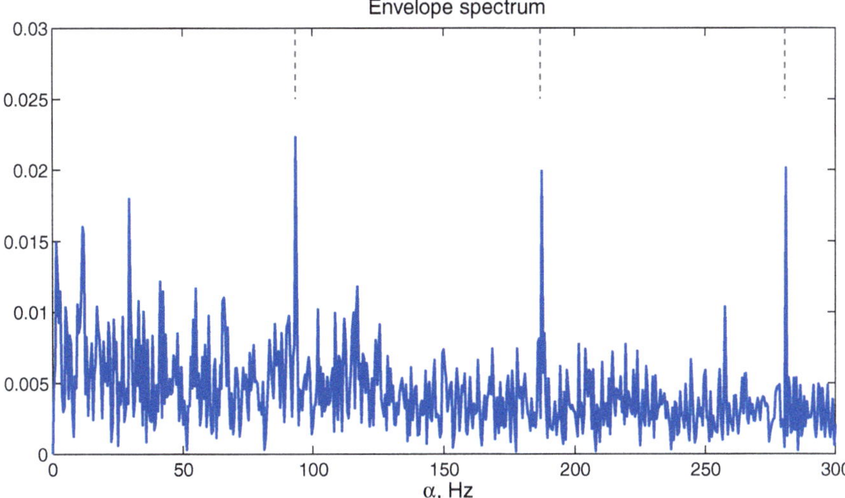

Fig. 3 Envelope spectrum for the faulty case computed from the high pass filtered version of the signal above 6 kHz. Top dashed lines indicate the defect frequency (BPFO) and harmonics

Consider now the SCoh of the faulty case, which is also shown in Fig. 4. Two important points are observed in this case. First, the defect is clearly identified by the presence of discrete components in the α axis, at the cyclic frequency of the defect (i.e. the BPFO) and harmonics, which are indicative of second-order CS related to the rolling of the bearing elements over the damaged zone. The next observation is that the second-order CS due to the defect appears in the frequency range between ≈ 14 and ≈ 25 kHz. This indicates that this is the optimum range for filtering the vibration signal previous to the construction of the envelope. Figure 5 shows the envelope spectrum obtained when filtering in this range. As expected, this results in a cleaner envelope spectrum, in which the defect is clearly identified. Note in this case it was not necessary to use the auto-covariance function for calculating the SCoh, because impure second-order CS components do not appear in the range in which the second-order CS related to the defect manifests.

At this point one could argue about the real benefits of using the SCoh, considering that the simpler method based in the comparison of the PSD resulted in an envelope spectrum in which the defect was, indeed, identified. In this respect, one should keep in mind, however, that the PSD-based methodology requires a previous measurement of the healthy case, which might not be available. Note how with the SCD approach the diagnosis can be effectively done without need of the the healthy signal, but only relying in the second-order CS features of the faulty signal.

Another alternative for the selection of the optimum filter, when only the faulty signal is available is the Kurtogram (Antoni 2007). The Kurtogram gives the combination of central frequency and bandwidth of the filter which maximizes the Spectral Kurtosis (SK). Figure 6 presents the Kurtogram of the faulty signal, which suggests

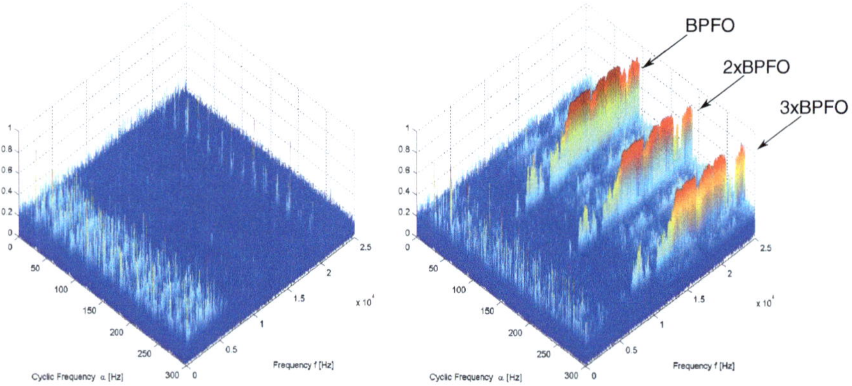

Fig. 4 Spectral coherence map for the non-faulty (*left*) and faulty case (*right*)

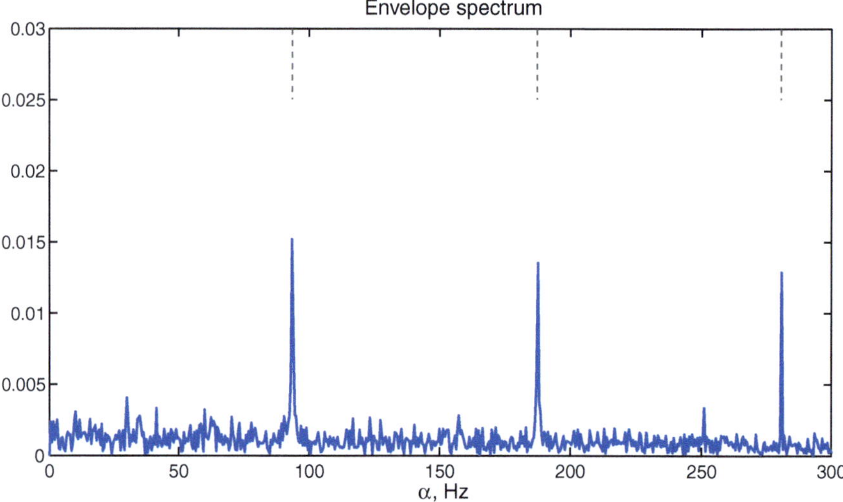

Fig. 5 Envelope spectrum for the faulty case computed from the high pass filtered version of the signal above 14 kHz. Top *dashed lines* indicate the defect frequency (BPFO) and harmonics

that the optimum filter is the high pass filter with cutoff frequency of ≈ 24 kHz. Figure 7 shows the envelope spectrum calculated using this information. Note the result is poorer when compared to the envelope spectrum of Fig. 5. This is because the range indicated by the Kurtogram is the range in which the maximum SK is obtained, which can occur due to some isolated impacts on the bearing instead of the repetitive train of impulses, or even due to some other process present in the signal with high SK. This can be observed in Fig. 8, where portions of the faulty signal filtered with the information obtained from the PSD, SCoh and Kurtogram are respectively presented. Their values of Kurtosis are 2.04, 83.00 and 90.71, respec-

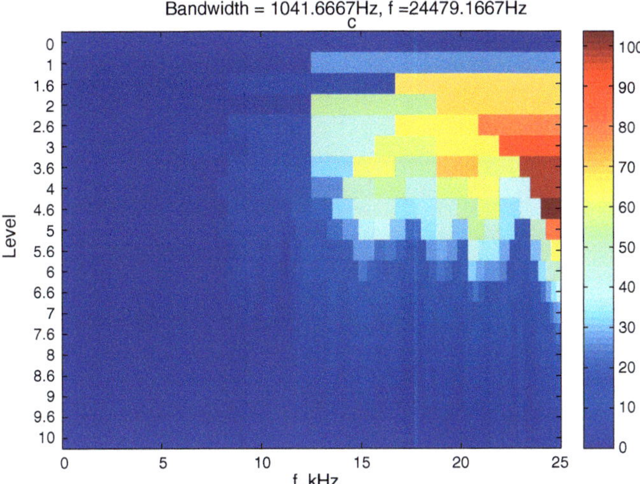

Fig. 6 Fast Kurtogram for the faulty case

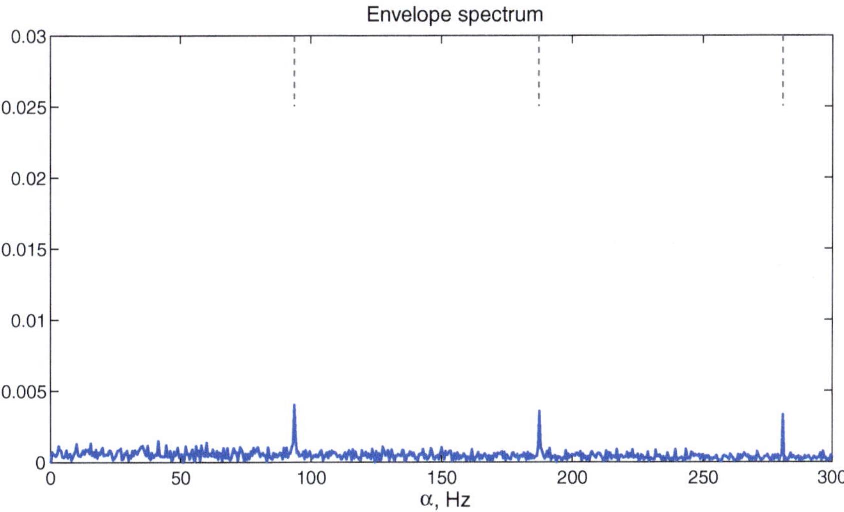

Fig. 7 Envelope spectrum for the faulty case computed from the high pass filtered version of the signal above 24 kHz. Top dashed lines indicate the defect frequency (BPFO) and harmonics

tively. As expected, the signal filtered with the information from the Kurtogram gives the maximum Kurtosis, but it is the signal filtered with the information from the SCoh map which better isolates the evidence of the defect. Note, however, that in this case the Kurtogram presents a frequency range in which the SK is not maximum, but still much higher than the rest, and that this range is in accordance with the range determined from the spectral coherence map.

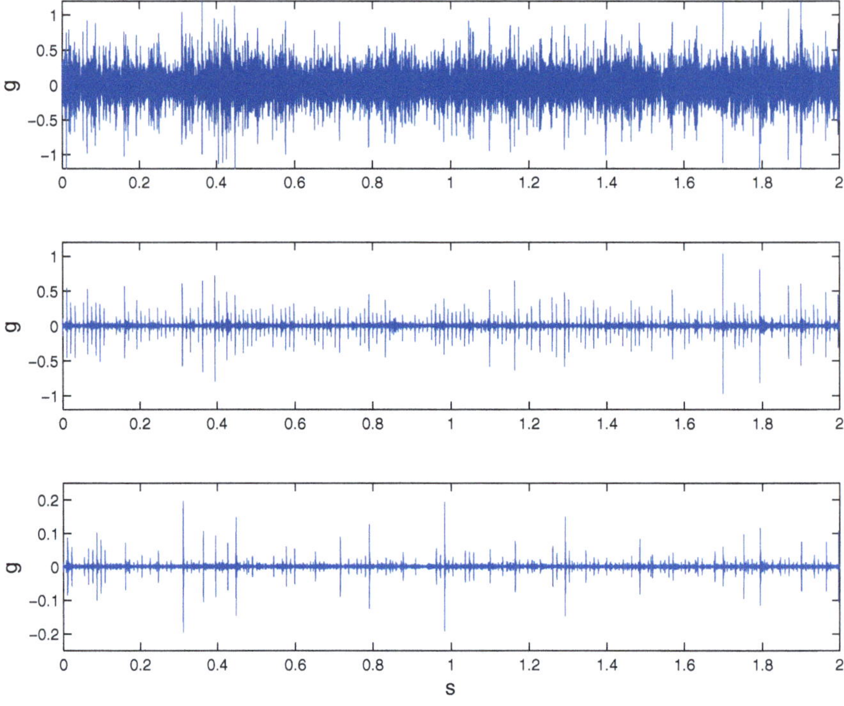

Fig. 8 Filtered versions of the vibration from faulty case. High-pass filter above 6 kHz (*top*), high-pass filter above 14 kHz (*center*) and high-pass filter above 24 kHz (*bottom*)

3.2 Case 2: Planetary Gearbox

This case relates to the AE measured on a planetary gearbox in two different conditions: non-faulty and with a localized defect on the inner ring of one planet bearing. Figure 9 shows the time history of the AE measured in both cases. In the non-faulty case, the signal presents a series of bursts spaced at the gear mesh period, although not in a strictly periodic way. The bursts from the gear meshing are amplitude-modulated due to the passing of the planet gears near the sensor position. As will be shown, this signal is CS of the second order, being the gear mesh frequency the only fundamental cyclic frequency contained in the signal.

In the faulty case, the signal also presents the bursts due to the gear meshing, although for an unknown reason in this case they are not as dominant as in the healthy case. Additionally, the signal contains repetitive bursts with second-order CS characteristics, due to the bearing fault. Since the bearing fault frequency is related to the gear mesh frequency in a non-commensurate fashion, the resulting signal is poly-cyclostationary at the second order.

Figure 10 shows the PSD of the faulty and non-faulty case. In this case, due to the difference in amplitude of the gear meshing AE, the non-faulty case presents

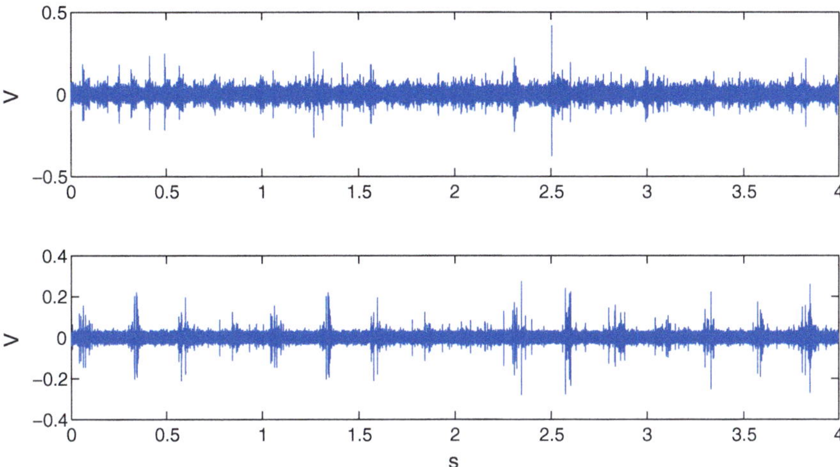

Fig. 9 Acoustic emissions measured on the healthy case (*top*) and on the faulty case (*bottom*)

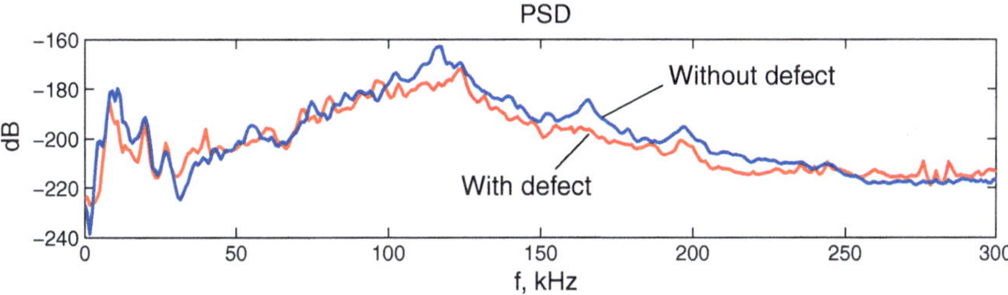

Fig. 10 Power spectral density of the faulty and non-faulty AE

higher magnitude in the PSD, so no valuable information can be obtained from these spectra. Both magnitude spectra (not shown) are broadband, presenting no lines at the gear mesh frequency and bearing fault frequency and harmonics, meaning the signals are non-stationary.

Figure 11 shows the envelope spectra calculated from both signals. No filter was considered, because the bursts appear to protrude significantly from the continuous part of the signal. As expected, they are dominated by the lines at the gear mesh frequency (288 Hz) and harmonics, being of higher magnitude in the non-faulty case. It is not possible to distinguish lines at the expected bearing fault frequency (89 Hz) and harmonics, nor sidebands around them spaced at the carrier rotating frequency (4 Hz).

Figure 12 shows the SCD map of the healthy and faulty case in the cyclic frequency range where the gear mesh frequency is contained. The appearence of the discrete components in the α-axis demonstrate the bursts generated by the gear meshing produce AE with second-order CS features. A similar picture is observed in the rest of the harmonics of the gear mesh frequency (not shown). Figure 13 presents the

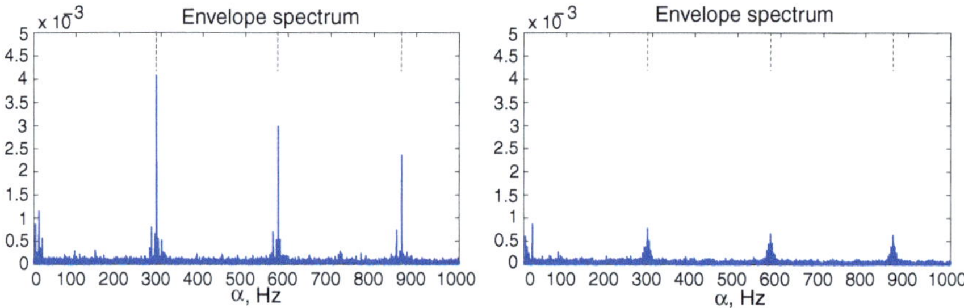

Fig. 11 Envelope spectrum for the healthy (*left*) and faulty (*right*) case, computed without filter. Top dashed lines indicate the gear mesh frequency and harmonics

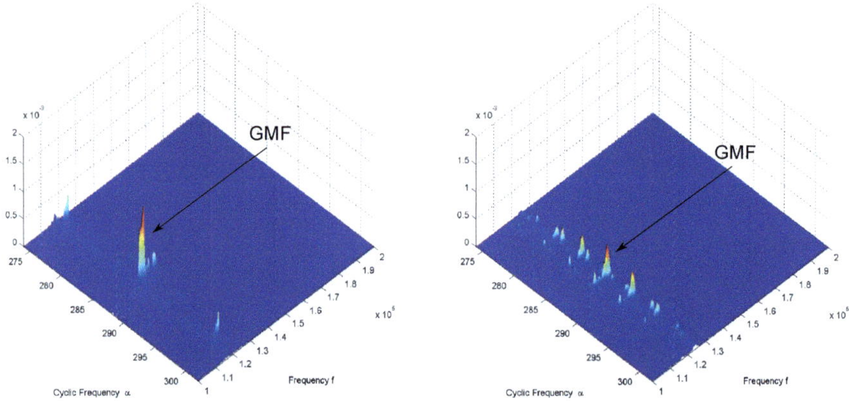

Fig. 12 Spectral correlation density map for the non-faulty (*left*) and faulty case (*right*), around the gear mesh cyclic frequency range

SCD map of both cases in the cyclic frequency range of the bearing fault frequency. The faulty case presents discrete components in the α-axis for the expected fault frequencies; whereas the healthy case does not. The presence of the discrete components in the α-axis at the gear mesh frequency and bearing fault frequency and harmonics, evidence the poly-CS characteristics of the AE in the faulty case. Note how in this case, from all the processing tools used, only the SCD map allowed the correct diagnosis of the fault.

4 Conclusions

There are a number of cases in which the behaviour of mechanical systems can be highly cyclostationary. The stationary assumption inherently made when analyzing these systems relying on the FFT of their measured variables is inappropriate in

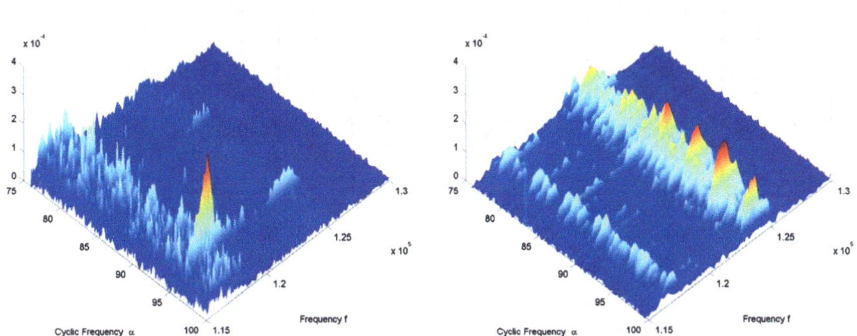

Fig. 13 Spectral correlation density map for the non-faulty (*left*) and faulty case (*right*), around the bearing fault cyclic frequency range

these cases. Advances in signal processing during the last years now permit the cyclostationary assessment of variables measured in mechanical systems without much effort. Notwithstanding, their analysis from the cyclostationary viewpoint still seems to be restricted to the scientific community. The reason for this situation probably relies in the difficulty found by practitioners on grasping the concepts related to cyclostationarity. It has been the aim of this work to contribute in this direction, by presenting the cyclostationary theory in a simple manner. The two cases presented show examples of cyclostationary behaviour of machines and how the analysis can be made under consideration of this fact.

References

Antoni J, Randall RB (2002) Differential diagnosis of gear and bearing faults. J Vib Acoust 24: 165–171

Antoni J, Daniere J, Guillet F (2002) Effective vibration analysis of IC engines using cyclostationarity. Part I-A methodology for condition monitoring. J Sound Vib 257:815–837

Antoni J, Bonnardot F et al (2004) Cyclostationary modelling of rotating machine vibration signals. Mech Syst Signal Process 18:1285–1314

Antoni J (2007) Cyclic spectral analysis in practice. Mech Syst Signal Process 21:597–630

Antoni J (2007) Fast computation of the kurtogram for the detection of transient faults. Mech Syst Signal Process 21:108–124

Antoni J (2009) Cyclostationarity by examples. Mech Syst Signal Process 23:987–1036

Capdessus C, Sidahmed M, Lacoume JL (2000) Cyclostationary processes: application in gear faults early diagnosis. Mech Syst Signal Process 14:371–385

Gardner W (1991) Exploitation of spectral redundancy in cyclostationary signals. IEEE SP Mag 8(2):14–36

Gardner W, Spooner C (1994) The cumulant theory of cyclostationary time-series. Part I foundation. IEEE Trans Signal Process 42:3387–3408

Ho D, Randall RB (2000) Optimisation of bearing diagnostic techniques using simulated and actual bearing fault signals. Mech Syst Signal Process 14:763–788

Raad A, Antoni J (2008) Indicators of cyclostationarity: theory and application to gear fault monitoring. Mech Syst Signal Process 22:574–587

Randall RB (2001) The relationship between spectral correlation and envelope analysis in the diagnosis of bearing faults and other cyclostationary machine signals. Mech Syst Signal Process 15:945–962

Spooner Chad M (1994) Higher-order statistics for nonlinear processing of cyclostationary signals. In: Gardner W (ed) Cyclostationarity in communications and signal processing. IEEE Press, New York

Spooner C, Gardner W (1994) The cumulant theory of cyclostationary time-series. Part II. Development and applications. IEEE Trans Signal Process 42:3409–3429

Vicuna Cristian M, Hoeweler C (2013) A method for extreme data reduction of acoustic emission (AE) data with application in machine failure diagnosis. In: Proceedings of surveillance 7 international conference, Chartres October 29–30

Model of the Planetary Gear Based on Multi-Body Method and Its Comparison with Experiment on the Basis of Gear Meshing Frequency and Sidebands

Dariusz Dąbrowski, Jan Adamczyk, Hector Plascencia Mora and Zahra Hashemiyan

Abstract Model tests can be used as a source of information about gearbox dynamics. Simulation of phenomena occurring in gearboxes allows to identify vibration signatures related to other failures of gears, bearings and shafts. In the chapter, the rigid-elastic model of the planetary gear is presented. The model was developed on the basis of the multi-body dynamics method. To conduct dynamic simulations specialized software MSC ADAMS was used. The multi-body method merges advantages of CAD modeling and efficient numerical simulations. Developed model allows to simulate the transmission error generated by a gearbox and forces in planet-sun and planet-ring meshing. In the tests it was observed that phase relations between meshing process influence on the sidebands of meshing frequency harmonics. The model was compared with experimental data on the basis of gears' meshing frequencies and modulation sidebands.

D. Dąbrowski (✉)
Department of Mechanics and Vibroacoustics, Faculty of Mechanical Engineering and Robotics, AGH University of Science and Technology, Al. A. Mickiewicza 30, 30-059 Krakow, Poland
e-mail: dabrowsk@agh.edu.pl

J. Adamczyk
Central Institute for Labor Protection-National Research Institute (CIOP-PIB), UL. Czerniakowska 16, 00-701 Warszawa, Poland
e-mail: adamczyk@agh.edu.pl

H. P. Mora
Department of Mechanical Engineering, Carretera Salamanca-Valle de Santiago km 3.5 + 1.8, Comunidad de Palo Blanco, University of Guanajuato, CP 36885 Salamanca, GTO, Mexico
e-mail: hplascencia@ugto.mx

Z. Hashemiyan
Faculty of Mechanical Engineering and Robotics, Department of Robotics and Mechatronics, AGH University of Science and Technology, Al. A. Mickiewicza 30, 30-059 Krakow, Poland
e-mail: zahra@agh.edu.pl

F. Chaari et al. (eds.), *Cyclostationarity: Theory and Methods*,
Lecture Notes in Mechanical Engineering, DOI: 10.1007/978-3-319-04187-2_11,
© Springer International Publishing Switzerland 2014

1 Introduction

A gearbox has a crucial influence on the reliability of a power transmission system. A gearbox is a complex object, which consist of shafts, gears, bearings and housing. The dynamic interactions that take place in a gearbox have a major effect on the vibrations and noise generated by the system. Identifying the dynamic phenomena that take place in gearboxes allows for a proper selection of the design, technological and operational features at the initial design stage (Łazarz and Peruń 2012).

Two main variants of gears dynamic models can be distinguished; models where all phenomena occurring in a power transmission system are included and models which take into account only phenomena inside the gearbox. In the first group, the dynamics of a motor, couples, gears and working machine are included. The second group considers physical phenomena occurring only inside the gearbox, and in these models the meshing stiffness and manufacturing of gears mostly affects the dynamics (Müller 1986, Bartelmus 1998). To model a gearbox dynamics it is possible to use lumped parameter models which are used for relatively simple gear systems (Åkerblom 2001). In the lumped parameter models the laws governing the system are described by differential equations. A vast review of mathematical models used in gear dynamics was presented by Özguven and Houser (1988).

It can be stated that the first dynamic model of gears was proposed in 1868 by Walker (1961). It was based on an empirical dynamic factor (DF). The DF was defined as a static load divided by a dynamic load. The first spring-mass model was introduced by Tuplin (1953), the meshing of two gears was modeled by a system with one-degree freedom. The model in which meshing stiffness varying in time was presented by Strauch (1953). In this model the changes in meshing stiffness were due to changing from a single pair to a double pair of teeth in the engage. In models examined later, the tooth stiffness was the main potential energy-storing element, as the other elements were rigid or neglected. The model developed by Bollinger and Bosh and discussed in Dąbrowski et al. 2000 is constructed form two masses representing gears connected by spring-damper element, the changes of meshing stiffness depends on number of teeth in engage and kinematic deviations were considered. More complex model was presented by Müller (1986) it allows to take into account a backlash and other kinds of geometrical deviations. Müller also introduced the dynamic model of planetary gear (Müller 1986). In this model the meshing stiffness, stiffness and dumping of bearings, meshing phase relations and non-linearity of phenomena occurring in kinematic pairs are considered. Another modeling approach is based on the apparent interface method and was presented by Radkowski at al. (1996), Filonik et al. (1998). This method is based on additional angular displacement of gears in relation to the actual displacement, which causes apparent overlap of the pinion and gear, the interference is compensated by elastic deformation of the teeth.

Łazarz et al. presented a model of the system consisting of a motor, shafts, gears and bearings (Łazarz and Peruń 2006, 2009). The presented models were verified on a test rig operating in the circulating power system. The presented results can be used for analysis of the methods, which can be used to reduce the vibroactivity of the gearboxes in an early design stage. In the study it was shown that it is important to analyze the design and the technological factors in gear modeling. The operation parameters, such as time-varying loading and speed, are crucial for a proper modeling of the phenomena in gearboxes. In the chapter (Bartelmus et al. 2010) models of fixed axis and planetary gearboxes, operating under varying load conditions were presented. In another study (Chaari et al. 2012), a lumped parameter model was applied to investigate the influence of meshing forces, variable loads and errors on the dynamics of gearboxes. The time varying operations and teeth faults were modeled by proper selection of the meshing stiffness function.

The modeling methods based on mathematical description of phenomena in gearboxes or multi-body dynamics approach are used increasingly and developed for modeling more complex gears systems (Inalpolat and Kahraman 2009; Viadero et al. 2012). The group of models, where a mathematical description of the observations is used instead of differential equations depicting physical processes are phenomenological models. A mathematical model describing the modulation mechanisms in a planetary gear was presented by Inalpolat et al. (2009). A vibration signal generated by the planetary gear is modeled on the basis of system parameters: the number of planets, planet position angles, and planet phasing relationships (Vicuña 2012). A relatively new approach for modeling the dynamics of gearboxes is based on the multi-body dynamics method. A multi-body system is a model of a real system built with the assumption that bodies in a real system are rigid or flexible and connected by joints. The multi-body system allows for time domain integration of the solution, which captures the non-linear effects of bearing stiffness and clearances, gear backlash, large rotations and other nonlinear phenomena (Palermo et al. 2010). Specialized multi-body dynamics software, such as ITI-SIM, SIMPACK, LMS Virtual. Lab Motion and MSC ADAMS, allow to model three-dimensional gear bodies, tooth micro-geometry, global and local tooth stiffness. The study describing the multi-body approach for modeling the vibration of gears was presented by Dresig et al. (2005). The model allows for dynamic simulations of planetary gearboxes, considering the stiffness characteristics. A new approach to modeling gear systems was presented by Ebrahimi and Eberhard (2006). It assumes that the teeth and the body of the gear wheels are rigid but are connected by elastic elements. The study related to a multi-body dynamics model developed in MSC ADAMS software were presented in few publications (Han et al. 2009; Kong et al. 2008; Viadero et al. 2012).

To sum up, lumped parameter models can be used for gear dynamics modeling or modeling of all gearboxes working in power transmission systems. The literature presenting this approach for modeling of gears dynamics is vast and still growing, in many cases models were positively verified in an experiment. The disadvantage of this approach lies in the modeling difficulty of complex systems. The phenomenological models allow for relatively simple modeling of vibrations generated by gears, but require detailed knowledge about all phenomena in gearboxes. Another method that is useful in modeling complex systems is multi-body dynamics method. It allows to model a dynamic system on the basis of the geometry of the elements and interacting forces.

2 Rigid-Elastic Model of the Planetary Gear

2.1 Multi-Body Method

A multi-body system is a model of the real system, built with the assumption that bodies in the real system are rigid or flexible and connected by joints, e.g. revolute, translational, cylindrical or spherical joints. The motion of elements is caused by other kinds of forces and torques. In the multi-body dynamics method mostly kinematic and dynamic analyses are carried out. The motion of a multi-body system is calculated by integration of differential equations. The configuration of a multi-body system is defined by a set of variables called generalized coordinates that completely define the location and orientation of each body in the system. Six independent coordinates (three coordinates describing the location and three the orientation) completely describe the configuration of a rigid body in the space (Shabana 2005). The generalized coordinates of the body reference $\mathbf{q}$ are given by the vector

$$\mathbf{q}^i = \left[\mathbf{R}^{iT}, \boldsymbol{\theta}^{iT} \right]^T, \tag{1}$$

where $\mathbf{R}$ is the coordinate vector of the origin of the body reference, $\boldsymbol{\theta}$ is the vector with Euler angles, and i represents number of rigid body. The vector describing the location and orientation of n bodies is given by

$$\mathbf{q} = \left[\mathbf{q}_1^T, \mathbf{q}_1^T, \mathbf{q}_1^T, \ldots, \mathbf{q}_n^T \right]^T. \tag{2}$$

To describe the multi-body system configuration in a space with n interconnected rigid bodies one needs $6n$ coordinates; these coordinates are not entirely independent because of the joints.

Derivation of the equations of motion for a multi-body system was particularly presented by Wojtyra and Frączek (2007). The motion of a multibody system can be described by set of following formulas

$$\left\{ \begin{array}{l} \mathbf{M}\dot{\mathbf{u}} - L_{\mathbf{R}}^T + \boldsymbol{\Phi}_{\mathbf{R}}^T \boldsymbol{\lambda} - \mathbf{H}_F^T \mathbf{F} = \mathbf{0}_{3nx1} \\ \dot{\mathbf{p}} - L_{\boldsymbol{\theta}}^T + \boldsymbol{\Phi}_{\boldsymbol{\theta}}^T \boldsymbol{\lambda} - \mathbf{H}_N^T \mathbf{N} = \mathbf{0}_{3nx1} \\ \mathbf{p} - L_{\varepsilon}^T = \mathbf{0}_{3nx1} \\ \mathbf{u} - \dot{\mathbf{R}} = \mathbf{0}_{3nx1} \\ \boldsymbol{\varepsilon} - \dot{\boldsymbol{\theta}} = \mathbf{0}_{3nx1} \end{array} \right\}, \tag{3}$$

where

$$\begin{aligned} \mathbf{u} &= \dot{\mathbf{R}}, \\ \boldsymbol{\varepsilon} &= \dot{\boldsymbol{\theta}}, \end{aligned} \tag{4}$$

In the equation above $\boldsymbol{\Phi}$ is the function describing constraints in a multibody system, L is the Lagrange function given by difference of kinetic and potential energy of a system, λ is the vector of Lagrange multipliers, $\mathbf{M}$ is the mass matrix, $\mathbf{p}$ is the generalized momentum vector, $\mathbf{F}$ and $\mathbf{N}$ are the vectors containing the external forces and torques acting on a multi-body system, and $\mathbf{H}^T{}_N$, $\mathbf{H}^T_F$ are the matrices that allow the conversion of vectors $\mathbf{F}$ and $\mathbf{N}$ to the generalized forces. Generalized coordinates have to satisfy the constraint given by (3), thus for this purpose Eq. (5) should be used (Wojtyra and Frączek 2007)

$$\mathbf{\Phi}(\mathbf{q}, t) = \mathbf{\Phi}(\mathbf{R}, \mathbf{\theta}, t) = \mathbf{0}_{mx1}. \tag{5}$$

Assuming that the forces and torques are described by the functions $\mathbf{f}$ and $\mathbf{n}$, vectors $\mathbf{F}$, $\mathbf{N}$ can be treated as variables vector

$$\begin{aligned} \mathbf{F} - \mathbf{f}(\mathbf{R}, \mathbf{\theta}, \mathbf{u}, \boldsymbol{\varepsilon}, \mathbf{F}, \mathbf{N}, t) &= \mathbf{0}_{3nx1}, \\ \mathbf{N} - \mathbf{n}(\mathbf{R}, \mathbf{\theta}, \mathbf{u}, \boldsymbol{\varepsilon}, \mathbf{F}, \mathbf{N}, t) &= \mathbf{0}_{3nx1}. \end{aligned} \tag{6}$$

The dynamics of a multi-body system is described by a set of Differential and Algebraic Equations (DAEs) (3) and (5–6). Simulations involve the solution of DAEs; there are two basic types of algorithms to perform the numerical integration; Stiff and Non-stiff solution methods. The Stiff solution methods use implicit Backward Difference Formulations (BDF) to solve the DAEs and the Non-stiff solution methods use explicit formulations to solve Ordinary Differential Equations (ODEs) that are obtained from the DAEs by way of coordinate partitioning methods (MSC ADAMS 2010).

2.2 Description of the Model

The gears dynamic models are mostly based on analysis of differential equations describing a system, the lumped parameter models are sometimes insufficient for modeling complex systems; so for these systems phenomenological models can be applied, but they require detailed knowledge about all phenomena in gearboxes. In the multi-body method the gear system is modeled by rigid bodies representing the gears' wheel, shafts and gearbox housing as well as elastic elements. The contacts are essential for modeling the vibrations generated by gears in the multi-body method. Properly defining the contact algorithm and parameters, such as stiffness and dumping, is crucial for gear mesh modeling. The multi-body method merges the advantages of CAD modeling and efficient numerical simulations (Dąbrowski et al. 2012, 2013; Dąbrowski and Adamczyk 2012).

In the chapter, the rigid-elastic model of the planetary gear developed on the basis of the multi-body method is presented. The model was built with following assumptions; bodies of the gears are rigid, contact surfaces are flexible, damping and stiffness in the contacts was assumed, ideal involute tooth geometry, gears are made from steel without any surface treatment and lubrication is neglected. In the study

Table 1 Basic gears parameters of planetary gear MERCURY 1-A

Parameter	Sun gear	Planet gear	Ring gear
Number of teeth	24	21	66
Outside diameter	55.3 mm	48.9 mm	136.2 mm
Pressure angle	20 deg.	20 deg.	20 deg.

firstly, 3D CAD model of one stage planetary gearbox MERCURY 1-A was built. The basic gears parameters are presented in Table. 1.

The geometry of the planetary gear was transferred into an ADAMS environment. In multi-body software the constraints, forces and contacts were modeled. The contact force defined in ADAMS is composed of two parts; the elastic component which acts like a nonlinear spring, and the damping force which is a function of the contact-collision velocity (Kong et al. 2008). The absolute value of the contact force can be expressed by following formula

$$F = \begin{cases} K(x_0 - x)^e + CS\dot{x} \text{ for } x < x_0 \\ 0 \text{ for } x \geq x_0 \end{cases}, \tag{7}$$

$$S = \begin{cases} 0 \text{ for } x > x_0 \\ (3 - 2\Delta d)\Delta d^2 \text{ for } x_0 - d < x < x_0 \\ 1 \text{ for } x \leq x_0 - d \end{cases}, \tag{8}$$

where d is the penetration depth, e is the contact force exponent, K is the contact stiffness and C is the damping coefficient. The contact stiffness between bodies, according to the Hertzian elastic contact theory, can be described by a pair of ideal contacted cylindrical bodies (Fisher 1961, Kong et al. 2008). The contact stiffness of a gear pair can be defined according to the expression

$$\begin{cases} K = \frac{4}{3}R^{1/2}E^* = \frac{4}{3}\left(\frac{id_i \cos(\alpha_t)\tan(\alpha_t')}{2(1+i)\cos(\beta_b)}\right)E^* \\ \frac{1}{E^*} = \frac{1-v_1^2}{E_1} + \frac{1-v_2^2}{E_2} \\ \beta_b = \text{atan}(\tan(\beta)\cos(\alpha_t)) \end{cases} \tag{9}$$

where R is the equivalent radius of two contacting bodies, E^* is the equivalent Young's modulus, E_1, E_2 is the Young's modulus for pinion and gear respectively, v_1 and v_2 are the Poisson ratio of pinion and gear, α_t' is the transverse pressure angle at engaged, α_t is the transverse pressure angle at standard pitch circle, β is the helical angle at the pitch and β_b is the helical angle at the base circle, i is the gear ratio, and d_1 is the diameter of the standard pitch circle.

In the study, it was assumed that the gears are made from steel with the Young modulus $2.1 \cdot 10^{11}$ Pa and the Poisson ratio 0.3. Contact stiffness calculated from Eq. (9) is equal to $K = 3 \cdot 105$ N/mm$^{3/2}$. Damping coefficient takes a value from 0.1 to 1 % of K, in the study it was assumed $C = 3{,}000$ Ns/mm, the force exponent $e = 1.2$ and the penetration depth $d = 0.3$ mm.

Fig. 1 Model of planetary gear **a** CAD model, **b** dynamic scheme of model

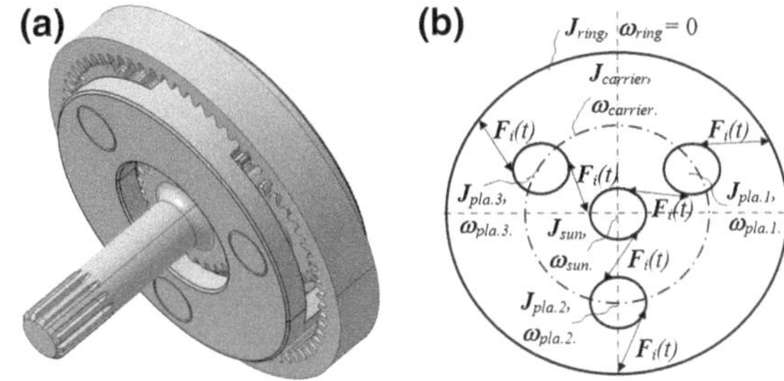

2.3 Analysis of Simulation Results

To conduct dynamic simulations of the model in the ADAMS following constraints were applied; fixed joint between the ring gear and ground, the revolute joints between the satellites and carrier and the carrier and ground. A constant rotation motion was applied to the sun gear and a constant torque to the carrier. The contacts were modeled between the planet and the sun gears, and the ring and the planet gears. The geometry of the planetary gear and dynamic scheme of models are presented in Fig. 1.

In simulations the integrator WSTIFF and the Stabilized Index-2 (SI2) formulation were used. The presented tests were conducted for a rotation frequency 27 Hz and torque 50 Nm.

To identify characteristic frequencies generated by the gearbox Gear Meshing Frequency (GMF) was calculated by following formula

$$f_{12} = f_{23} = f\frac{z_1 z_3}{z_1 + z_3} = 17.6 f_1, \tag{10}$$

where f_{12} and f_{23} are the meshing frequency for the sun and the planet gear respectively, f_1 is the rotational frequency for the input shaft. The modulation frequency in a planetary gear caused by passing of planet gears trough constant point on a gearbox housing is given by the following relation

$$f_m = f_a s = 0.8 f_1, \tag{11}$$

where

$$f_a = f_1 \frac{z_1}{z_1 + z_3} = 0.266 f_1, \tag{12}$$

In equation above f_a is the carrier rotational frequency, $s = 3$ is the number of planet gears. Vibrations generated by a gearbox are mostly related to the Transmission Error (TE). The transmission error is defined as

$$TE(t) = \dot{\theta}_1(t) - i\dot{\theta}_2(t), \tag{13}$$

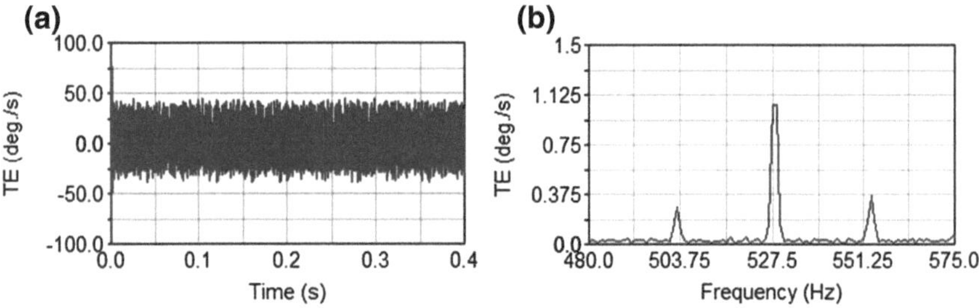

Fig. 2 Transmission error generated by model of planetary gear, **a** time signal, **b** spectrum

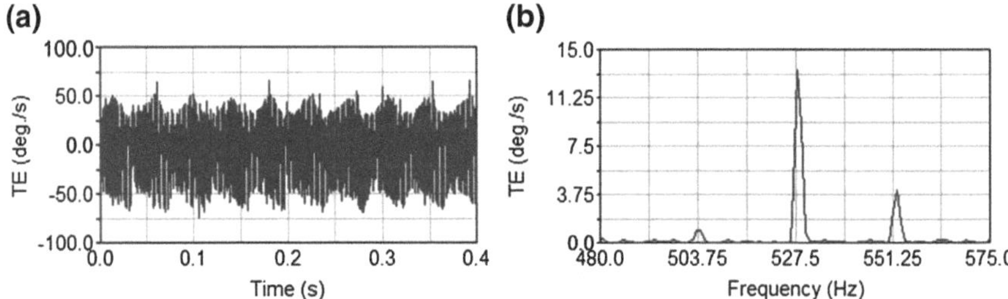

Fig. 3 Transmission error generated by model of planetary gear for misalignment of sun gear, **a** time signal, **b** spectrum

where θ_1 is the input rotational speed, θ_2 is the output rotational speed, i is the gear ratio. The TE represents all interactions occurring in a gearbox system, therefore in the tests the transmission error generated by the model of the planetary gear was analyzed.

On the spectrum presented in Fig. 2b one can observe the excitation on frequency 528 Hz and the sidebands. The frequency 528 Hz is related to the meshing frequency $f_{12} = f_{23}$, and the sidebands are caused by the modulation $f_m = 24$ Hz. The examined planetary gear belongs to the group of planetary gears with equally-spaced planet gears and in-phase gear meshing processes, because the angle between the positions of the planet gears is equal to 120 deg., and all meshes are in-phase, which means that phase difference between vibration generated in the meshing between the i-th planet gear and the ring gear, as experienced by an observer standing in the carrier plate must be either zero or a positive integer, this requires z3/s to be a positive integer, for the examined case it is equal to 33 (Vicuña 2012). That's why on the spectrum (Fig. 2b) occur symmetric sidebands with frequency f_m.

The most common fault of gearboxes working in the industry is misalignment (parallel or angular), which can be caused by manufacturing errors as well as bearing defects. In the next tests the misalignment of the sun gear was simulated; it was introduced by a vertical and horizontal shift of the sun gear axis about 0.1 mm and its rotation about 0.5 deg. in x-y plane. In Fig. 3 the transmission error generated by the model with modified geometry and its spectral analysis are presented.

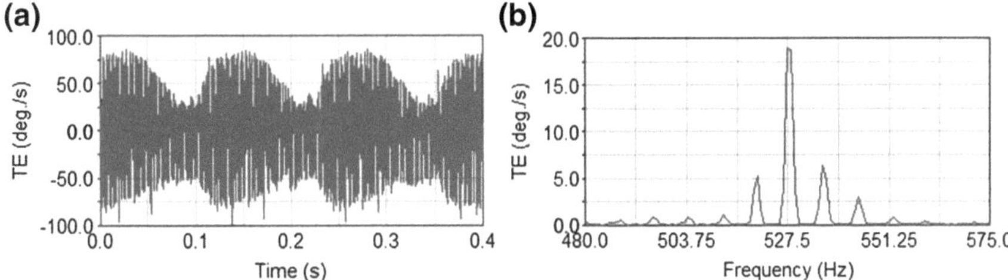

Fig. 4 Transmission error generated by model of planetary gear for misalignment of sun and planet gear, **a** time signal, **b** spectrum

After modification of the assembly the signal of the TE changes character, and the GMF increases about 10 times; also, amplitude of the sidebands changes, as the right sideband has a bigger amplitude than the left one. The misalignment of gears result in changes of the load distribution of a gear pair that results in increasing contacts and bending stresses, this phenomena reveal in the spectrum by increase of second GMF and sidebands of meshing frequencies (Randall 2011). The changes of the sidebands are related to the structure modification, after introduction of the misalignment, there is no in-phase meshing process between the sun and planet gears.

In the last tests additionally the misalignment of one planet gear were introduced to the model; the planet gear axis was rotated about 0.2 deg. in x-y plane.

The TE for this model (Fig. 4a) has a two times bigger peak-to-peak value than for the model without modifications (Fig. 2a); also, the cyclostationary character of this signal can be observed. The structure of the spectrum for this case (Fig. 4b) is similar to the spectrum of signals from the group of planetary gearboxes with unequally-spaced planet gears and out-of-phase meshing processes (Vicuña 2012); the sidebands are related to the carrier rotational frequency of 8 Hz and its multiples.

3 Comparison of the Model Tests with the Experiment

3.1 The Experiment

The experiment was conducted on the test rig with a planetary gearbox, designed to perform simulations under varying operations. The test rig consists of the asynchronous motor, one stage planetary gearbox MERCURY 1-A with gear ratio of 3.75, and electromagnetic particle break. The motor and planetary gearbox are coupled by an elastic coupling, in turn the planetary gearbox and break are coupled by a rigid coupling. The Modbus communication protocol allows to control all of the motor parameters, the torque is controlled by application of the electromagnetic particle break EMA-ELFA.

Fig. 5 Measuring system

Table 2 Equipment used in experiment

Number	Name
1	Accelerometer 356A15
2	Laser Tacho Probe MM0360
3	NI sbRIO 9602 Controller
4	NI cRIO 9233 Module

The experiment was conducted in the Laboratory of Mechanical Diagnostics at the AGH University of Science and Technology in Krakow. Figure 5 presents the measuring system.

In the experiment acceleration and keyphasor signals were measured by the sbRIO Controller with cRIO 9233 module. The PC computer was used for a data logging by the TCP/IP protocol. The equipment used in the experiment is listed in Table 2.

The acceleration signal was measured on the gearbox housing and the keyphasor signal was measured on the gearbox input shaft. The measurements were conducted with sampling frequency of 5 kHz. The experiment was conducted in the steady-state conditions for rotational speed of 27.7 Hz and torque equal to 50 Nm.

3.2 Comparison of the Results

To compare the results from the model tests and experiment the signals were analyzed in the time and frequency domain. In the experiment the vibration signal was measured on the gearbox housing, but in the model tests the TE was registered. Figure 6 presents comparison of the signals from the model tests and the experiment.

The cyclostationary character of the analyzed signals is noticeable (Fig. 6a, b). In Fig. 6c, d the gear meshing frequency harmonics are visible, the location of the GMFs is related to the geometry of the PG and operation parameters. In the experiment and

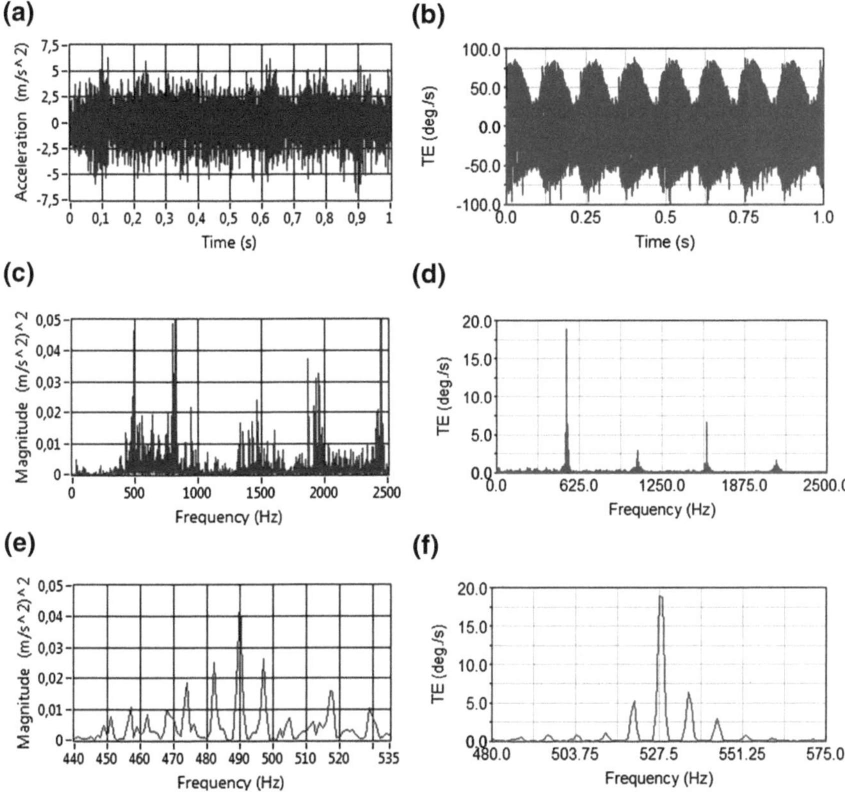

Fig. 6 Comparison of signals from model tests and experiment, **a** acceleration signal from experiment, **b** TE from model tests, **c** power spectrum of acceleration signal from experiment, **d** spectrum of TE from model tests, **e** power spectrum (GMF band) of acceleration signal from experiment, **f** spectrum (GMF band) of TE for model

model tests it was observed that the first GMF has the biggest amplitude, and the odd harmonics are more significant. In the frequency band related to the first GMF (Fig. 6e, f), one can observe that the main frequency is related to the meshing process (in experiment 489 Hz and model test 528 Hz) and the sidebands which depend on the phase relations between the meshing process of the planet gears.

4 Conclusion

The developed model of the planetary gear allows for simulations of the transmission error and meshing force generated by the planet-sun and planet-ring meshing process. The gear bodies are connected in meshing points by force elements, which were modeled on the basis of the contact algorithm. The contact force is a non-linear function of deformation depth and it depends on the stiffness and damping parameters. A planetary gearbox is a complex object, where vibration signals are generated by other elements interacting with each other, and they are transmitted by other paths to a signal receiving point. In the presented study the transmission error generated by the model of the planetary gear consists of a basic gear meshing frequency harmonic and sidebands. After introduction of manufacturing error (misalignment) the meshing phase relations for the planet gears change; the spectrum of the transmission error is similar to the spectrum of the vibrations generated by the planetary gears with unequally-spaced planet gears and out-of-phase meshing processes.

Summarizing, on the basis of a preliminary verification, it can be stated that the presented rigid-elastic model of the planetary gear allows to simulate phenomena related to the meshing process. It should be noted that the presented study is preliminary and that the proposed model has to be verified in detail with the object. In the future the test rig will be equipped with high resolution encoders which will allow to measure the transmission error; also, the model will be developed by flexible bodies that will allow to model the deflection of teeth and shafts, and taking into account Coulomb friction and lubrication.

Acknowledgments The authors gratefully acknowledge the helpful comments and suggestions of the reviewers, which have improved the presentation.

References

Åkerblom M (2001) Gear noise and vibration–a literature survey. Volvo construction equipment components AB. http://www.diva-portal.org/smash/get/diva2:139878/FULLTEXT01. Accessed 15 July 2013

Bartelmus W (1998) Diagnostyka maszyn górniczych górnictwo odkrywkowe. Wydawnictwo "Śląsk", Katowice

Bartelmus W, Chaari F, Zimroz R, Haddar M (2010) Modelling of gearbox dynamics under time-varying nonstationary load for distributed fault detection and diagnosis. Euro J Mech A Solid 29:637–646

Dąbrowski D, Adamczyk J (2012) Analysis of vibrations generated by the multi-body model of a planetary gear. Mech Control 31(4):143–149

Dąbrowski D, Adamczyk J, Mora PH (2012) A multi-body model of gears for simulation of vibration signals for gears misalignment. Diagnostyka-Appli Struct Health Usage Condition Monitor 62(2):15–23

Dąbrowski D, Batko W, Cioch W (2013) Application of laser measurements for validation of the gears model with an experimental data. Acta Physica Polonica A 123(6):1016–1019. doi:10.12693/APhysPolA.123.1016

Dąbrowski Z, Radkowski S, Wilk A (2000) Dynamika przekładni zębatych. Wydawnictwo i Zakład Poligrafii Instytutu Technologii i Eksploatacji, Warszawa

Dresig H, Schreiber U (2005) Vibration analysis for planetary gears. Modeling and multibody simulation. In: International conference on mechanical engineering and mechanics, Nanjing, China, 26–28 October 2005

Ebrahimi S, Eberhard P (2006) Rigid-elastic modeling of meshing gear wheels in multibody systems. Multibody Syst Dyn 16(1):55–71

Chaari F, Zimroz R, Bartelmus W, Fakhfakh T, Haddar M (2012) Model based investigation on a two stage gearbox dynamics under non-stationary operations. In: Proceedings of the 2nd international conference CMMNO'2012. Springer, Tunisia, pp 133–143

Filonik R, Mączak J, Radkowski S (1998) Apparent interface method as a way of modelling the meshing process disturbances. Mach Dyn Prob 19:95–108

Fisher A (1961) Factors in calculating the load-carrying capacity of helical gears. Machinery 98:545–552

Han B, Cho M, Kim C, Lim C, Kim J (2009) Prediction of vibrating forces on meshing gears for a gear rattle using a new multi-body dynamic model. Int J Autom Technol 4(10):469–474

Inalpolat M, Kahraman A (2009) A theoretical and experimental investigation of modulation sidebands of planetary gear sets. J Sound Vib 323:677–696

Kong D, Meagher J M, Xu C, Wu X, Wu Y (2008) Nonlinear contact analysis of gear teeth for malfunction diagnostics. In: IMAC XXVI conference and exposition on structural dynamics, Orlando-Florida, 4 February 2008

Łazarz B, Peruń G (2006) Wpływ uszkodzeń elementów przekładni obiegowej na siły w zazębieniach. Transp Prob 1:23–38

Łazarz B, Peruń G (2009) Identification and verification of simulation model of gears working in circulating power system. Diagnostyka 52(4):55–60

Łazarz B, Peruń G (2012) Influence of construction factors on the vibrational activity of the gearing. Transp Prob 7(2):95–102

MSC Adams (2010) Reference manual

Müller L (1986) Przekładnie zębate dynamika. Wydawnictwo Naukowo-Techniczne, Warszaw

Nevazt Ozguvent H, Houser DR (1988) Mathematical models used in gear dynamics-a review. J Sound Vib 121(3):383–411

Palermo A, Mundo D, Lentini A, Hadjit R, Mas P, Desmat W (2010) Gear noise evaluation through multibody te-based simulations. In: Proceedings of ISMA Leuven, Belgio, pp 3033–3046

Radkowski S (1996) Zmiana podatności zębów jako parametr diagnostyczny. Przegląd Mechaniczny 5–6:15–48

Randall R (2011) Vibration-based condition monitoring industrial, areospace and automative applications, 1st edn. Wiley, Chichester

Shabana A (2005) Dynamics of multibody systems, 3rd edn. Cambridge University Press, Chicago

Strauch H (1953) Zahnradschwingungengungen (Gear vibrations). Zeitschrift des Vereines Deutscher Ingiere 95:159–163

Tuplin WA (1953) Dynamic loads on gear teeth. Mach Des 25:203–211

Viadero F, Fernandez del Rincon A, Liano E (2012) Dynamic analysis of an offshore wind turbine drivetrain on a floating support. In: Proceedings of the 2nd international conference CMMNO'2012. Springer, Tunisia, pp 627–634

Vicuña M (2012) Theoretical frequency analysis of vibrations from planetary gearboxes. Forsch Ingenieurwes 76:15–31

Wojtyra M, Frączek J, (2007) Metoda układów wieloczłonowych w dynamice mechanizmów. Oficyna Wydawnicza Politechniki Warszawskiej, Warszawa

Periodic Autoregressive Modeling of Vibration Time Series From Planetary Gearbox Used in Bucket Wheel Excavator

Agnieszka Wyłomańska, Jakub Obuchowski, Radosław Zimroz and Harry Hurd

Abstract Vibration signals acquired from machines operating under non-stationary operations are difficult to process due to their time varying spectral content, statistical properties, signal to noise ratio etc. In case of damaged machine vibration analysis, the classical damage detection approach might be defined as informative and non-informative contents separation. It can be done in many ways, including model based approaches. One of the most known solutions for constant load/speed operations exploits autoregressive (AR) modeling of the deterministic high energy components that often mask the weak impulsive and stochastic part of the signal. After establishing the model, the residual signal is extracted and further analyzed. In the case presented here, AR modeling is considered inappropriate because of the variation of speed/load conditions. To illustrate importance of the problem and novelty of our approach, a planetary gearbox vibration will be analyzed. The gearbox operates in a bucket wheel excavator (heavy duty mining machine) subjected to cyclic load/speed variation due to the digging/excavating process. Due to periodicity of the excavation process, it seems appropriate to assume a periodic autoregressive (PAR) model for the deterministic high energy components. In the chapter several topics will be discussed: real data inspired motivation for PAR modeling, estimation

A. Wyłomańska (✉)
Hugo Steinhaus Center, Institute of Mathematics and Computer Science, Wroclaw University of Technology, Janiszewskiego 14a, Wroclaw, Poland
e-mail: agnieszka.wylomanska@pwr.wroc.pl

J. Obuchowski · R. Zimroz
Diagnostics and Vibro-Acoustics Science Laboratory, Wroclaw University of Technology, Na Grobli 15, Wroclaw, Poland
e-mail: jakub.obuchowski@pwr.wroc.pl

R. Zimroz
e-mail: radoslaw.zimroz@pwr.wroc.pl

H. Hurd
Department of Statistics, The University of North Carolina at Chapel Hill, Chapel Hill, NC, USA
e-mail: hurd@stat.unc.edu

F. Chaari et al. (eds.), *Cyclostationarity: Theory and Methods*,
Lecture Notes in Mechanical Engineering, DOI: 10.1007/978-3-319-04187-2_12,
© Springer International Publishing Switzerland 2014

details, simulations and PAR based inverse filtering for extraction of the informative stochastic part of the signal. Finally, we present some comparison of PAR and AR for modeling the deterministic high energy part.

1 Introduction

Vibration signals acquired from complex mechanical systems (helicopters, wind turbines, mining machines, etc. (Bartelmus and Zimroz 2009; Samuel and Pines 2005; Urbanek et al. 2013)) usually require advanced signal processing for damage detection, especially when considering so called early damage detection.

Special and relatively challenging situation for vibration based condition monitoring are time varying operation conditions, i.e. varying load of machine and associated variation of rotating element speed. It may happen that variation of operating condition is in fact switching from regime to regime. This simple case may well be treated as two separated cases with constant load/speed. Another approach initiated in the last decade is the utilization of the instantaneous variation of operating conditions. It might be concluded that vibration signals from machine operating under non-stationary conditions are quite difficult to process due to their time varying spectral content, statistical properties, signal to noise ratio etc. (Bartelmus and Zimroz 2009; Combet and Zimroz 2009). A bucket wheel case discussed in this chapter can be a really good illustration of this problem.

An important idea in damage detection by analysis of machine vibration signals is the separation of the informative and non-informative parts of the signal. Basically, the Signal of Interest (SOI) that carries information about the presence of faults might be defined as low energy, random signal with amplitude modulation, or simpler as cyclostationary signal (of order 2) (Antoni 2005; Antoni et al. 2004; Makowski and Zimroz 2013; Obuchowski et al. 2013; Randall and Antoni 2011; Urbanek et al. 2013; Zimroz and Bartelmus 2012). The non-informative part of the signal is the deterministic high energy part and is the part we wish to remove or suppress by advanced signal processing methods.

This separation can be realized in many ways, including model based approaches (Baillie and Mathew 1996; Endo and Randall 2007; Makowski and Zimroz 2011; Poulimenos and Fassois 2006; Wang and Wong 2002; Zhan et al. 2006; Zhan and Mechefske 2007). One of the most known solutions for constant load/speed operations exploits autoregressive (AR) modeling for deterministic high energy components (Baillie et al. 1996; Endo et al. 2007; Makowski et al. 2013; Wang et al. 2002 and many others). Typical methodology is: after establishing the model and model based inverse filter, the residual signal (informative part) is extracted and further analyzed.

In the case presented here, AR modeling due to variation of speed/load conditions (affecting spectral contents) is not appropriate or certainly not optimal (Makowski and Zimroz 2011). To illustrate importance of the problem and novelty of our approach, a planetary gearbox vibration will be analyzed. The gearbox operates in bucket wheel excavator, a heavy duty mining machine subjected to cyclic load/speed variation due to digging/excavating process. To model such a data, it is proposed to use special case of AR model with time-varying coefficients (Makowski and Zimroz 2013; Poulimenos and Fassois 2006) and more specifically, due to periodicity of excavation process affecting the gearbox vibration, a periodic autoregressive model (PAR) is strongly suggested. In the chapter several topics will be discussed: real data inspired to PAR modeling, estimation details, simulations and PAR based inverse filtering for informative signal extraction.

2 Periodic Autoregressive Model

2.1 Basic Definitions and Properties

Definition 1. (Gladyshev 1961) The periodic autoregressive time series $\{X(t)\}$ of order p is defined as follows:

$$X(t) - \sum_{i=1}^{p} a_i(t)X(t - i) = b(t)Z(t), \tag{1}$$

where $\{Z(t)\}$ is a white noise time series and the coefficients $\{a_i(t)\}\, i = 1, 2, \ldots, p$, $\{b(t)\}$ are periodic with the same period T. Usually it is assumed the time series $\{Z(t)\}$ is a white noise.

The periodic autoregressive time series (PAR) is a special case of PARMA sequence (periodic autoregressive moving average), i.e. a time series which is defined as:

$$X(t) - \sum_{i=1}^{p} a_i(t)X(t - i) = b_0(t)Z(t) + \sum_{j=1}^{q} b_j(t)Z(t - i). \tag{2}$$

In the above definition the sequences coefficients $\{a_i(t)\}\ \ i = 1, 2, \ldots, p$, $\{b_j(t)\}$ $j = 0, 1, \ldots, q$ are also periodic with the same period T and the series $\{Z(t)\}$ is a white noise.

The PARMA sequence is one of the main time series which can be used as a model for periodically correlated (or cyclostationary) processes. A sufficient condition for a PARMA sequence $X(t)$ to be periodically correlated with period T can be obtained through re-expressing (2) in terms of the vector sequence $X(t)$ formed by blocking $X(t)$ into successive vectors of length T. Then (2) becomes

$$\varphi(B)X(n) = \theta(B)\varepsilon(n)$$

where $\varphi(B)$ and $\theta(B)$ are easily found from the parameters $\{a_i(t)\}, i = 1, 2, \ldots, p$, $\{b_j(t)\}\ \ j = 0, 1, \ldots, q\ \ t = 0, 1, \ldots, T - 1$. The PARMA sequence $X(n)$ will be periodically correlated if and only if $X(n)$ is stationary and a sufficient condition for this is

$$\det[\varphi(z)] \neq 0 \quad for \ |z| \leq 1.$$

Generally, periodically correlated (PC) random processes of second order are random systems in which there exists a periodic rhythm in the structure that is generally more complicated than periodicity in the mean function (Hurd and Miamee 2007). The exact definition of PC processes is as follows:

Definition 2. (Gladyshev 1961) A second order process $\{X(t)\}_{t \in Z}$ is called periodically correlated with period T if for every $s, t \in Z$ the following conditions hold:

$$m(t) = E(X(t)) = m(t + T)$$
$$R(s, t) = Cov(X(t), X(s)) = R(s + T, t + T).$$

And there are no smaller values of $T > 0$ for which the above conditions hold.

Due to their interesting properties periodically correlated time series have received much attention in the literature because they provide, for periodically non stationary phenomena an alternative to the conventional stationary time series. Examples occur in hydrology (Vecchia 1985) meteorology (Bloomfield et al. 1994), economics (Broszkiewicz-Suwaj et al. 2004; Parzen and Pagano 1979) and electrical engineering (Gardner and Franks 1975).

The PARMA system was also considered in case of infinite variance, see (Nowicka-Zagrajek and Wyłomańska 2006). In this case the covariance function cannot be considered as a measure of dependence therefore the cyclostationarity is expressed in the language of other measures of dependence, like codifference. For more details please see (Nowicka-Zagrajek and Wyłomańska 2006).

2.2 Estimation

The PAR processes are more common in practice than the general class of PARMA sequences therefore in this section we focus on the estimation procedure for PAR time series.

There are many methods that can be used to estimation of PAR coefficients. One of the methods is so called Yule-Walker method which is a consequence of method of moments. This method is very often used in practice because of the simple form of estimators. Let us mention the method can be used in case $p < T$.

Let us assume the random sample $X(1), X(2), \ldots, X(NT)$ comes from the PAR model with period T and order p. In practice, the parameters T and p may not be known but can be estimated by using statistical methods; for estimation of T, see (Hurd and Miamee 2007) and for p, a common method is to use the AIC or BIC criteria (Brockwell and Davis 2006). In our procedure we assume the period and order of the PAR model are known. In the first step of the analysis we compute the empirical (or sample) periodic mean by:

$$\hat{m}(v) = \frac{1}{N} \sum_{n=0}^{T-1} X(nT + v), \qquad v = 1, 2, \ldots, T$$

Next we subtract the empirical periodic mean to form:

$$Y(nT + v) = X(nT + v) - \hat{m}(v), \qquad n = 1, 2, \ldots, N, v = 1, 2, \ldots, T$$

The sequence $Y(1), Y(2), \ldots, Y(NT)$ constitutes also a realization of PAR sequence (1) therefore it satisfies the following equation:

$$Y(nT + v) - \sum_{i=1}^{p} a_i(nT + v)Y(nT + v - i) = b(nT + v)Z(nT + v).$$

Now multiplying Eq. (2) by $Y(nT + v - i)$ for $i = 0, 1, \ldots, p$ and taking the expected value we obtain $p + 1$ equations which can be written in a matrix form:

$$\Gamma_v A_v = R_v, \qquad v = 1, 2, \ldots, T$$
$$b_v^2 = R(v, v) - A_v' R_v,$$

where Γ_v is a square matrix which is defined as follows:

$$(\Gamma_v)_{ij} = R(v - i, v - j) = EY(v - i)Y(v - j), \qquad i, j = 1, 2, \ldots, p$$

Moreover

$$A_v = [a_1(v), \ldots, a_p(v)]'$$
$$R_v = [R(v, v - 1), R(v, v - 2), \ldots, R(v, v - p)]'.$$

The estimators of the parameters A_v and b_v we can obtain by replacing the theoretical covariances by the empirical ones according to the following:

$$\hat{R}(k, l) = \frac{1}{N} \sum_{n=0}^{N-1} Y(nT + k)Y(nT + l)$$

The estimators calculated by using the Yule-Walker method under the assumption of Gaussian distribution of the residual series $\{Z(t)\}$ are consistent and have asymptotically Gaussian distribution.

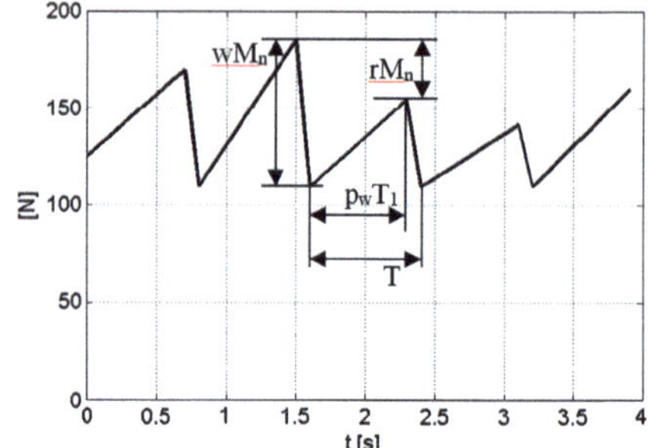

Fig. 1 Theoretical variation of external load in bucket wheel excavator during digging

3 Vibration Data Understanding and Simulation

3.1 Real Data Description

The BWE is one of the largest machines in mining industry. A precise description of its structure and design probably requires another chapter. From our perspective, it is important to say that our object is a multistage gearbox (with planetary gear-set inside) used in drive unit for driving bucket wheel (see Fig. 1, left panel). Due to digging process (that can be considered as almost periodic) and many factors with random nature (geology, human factors,...), the external load subjected to drive unit is time varying and looks like sawtooth time series with random amplitudes (strong variation) and period (relatively small variation), see Fig. 1, right panel.

As a result, the vibration signal acquired from the gearbox operating under such conditions has time varying (according to load variation) structure, Figs. 2, 3 (Zimroz et al. 2011; Chaari et al. 2012). Main spectral components, related to meshing phenomena or shaft rotation are frequency modulated with respect to engine output shaft speed variation and amplitude modulated according to engine torque variation. If one considers BWE and cyclicity of digging process, it might be concluded that discrete components in the spectrum are periodically varying. Figure 2 shows spectrogram of real vibration data example: 3 components related to meshing phenomena in BWE planetary gearbox presented as horizontal line can be characterized as FM modulated components. Moreover, one of line (in the middle) clearly indicates (marked by arrow) cyclic amplitude modulation with the same period (varying intensity of the spectrogram color)It is obvious that AR model with constant coefficients cannot be used here. AR model with constant coefficient is not able to track variation of components frequencies. Estimation error of AR model will be proportional to FM modulation index.

Purpose of this work is to apply PAR model to identify discrete components, build model-based inverse filter and filter the signal to extract signal of interest. What is signal of interest here? Structure of the signal showed in Fig. 2 is not complete.

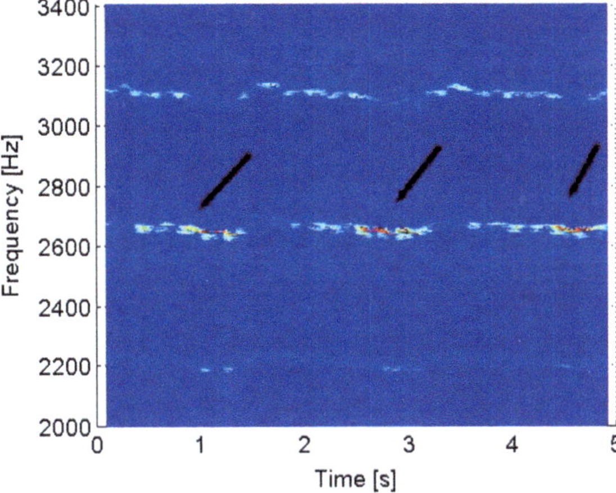

Fig. 2 Example of time frequency map of planetary gearbox vibration under time varying cyclic load

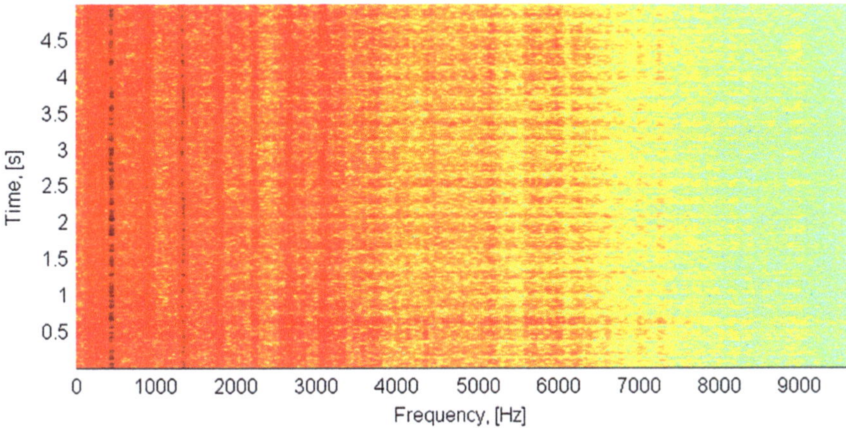

Fig. 3 Example of time frequency map of planetary gearbox vibration under time varying cyclic load with local damage (*vertical lines*—components to be modeled, *horizontal*—wideband excitation related to SOI we want to extract)

In case of local damage, apart from set of discrete (narrowband) components, one will notice cyclic wideband excitation related to cyclic impulsive disturbance in time signal (Fig. 3). Autoregressive model is not suitable to describe such phenomena, mentioned wideband excitation will not be included in the model.

During simulation (exciting the AR/PAR model by white noise) horizontal lines will not appear. Model based inverse filter can be interpreted as set of band stop filters and will "dump" the signal at mentioned discrete frequencies and "pass" rest of the signal. In this way, extraction of SOI is possible.

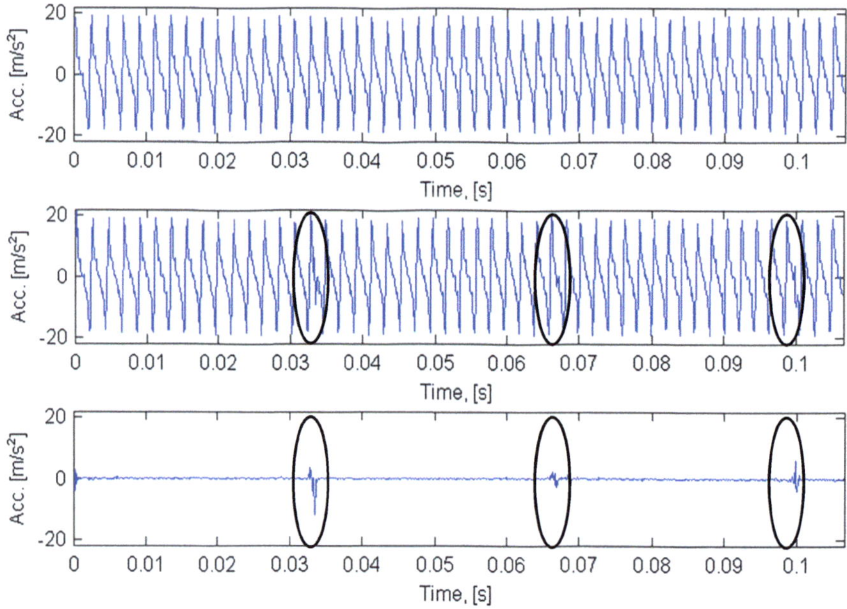

Fig. 4 Time series of healthy (*top panel*) and faulty (*center panel*) signal and SOI of faulty signal (*bottom panel*). Three places suspected of local damage occurrence are highlighted

In order to model vibration signal generated by gearbox operating under time varying conditions (in particular its frequency contents) and use this model for SOI extraction, we propose to start with simpler synthetic signal to prove efficiency of our method.

3.2 Signal Simulation

In this section we present a simulated signal that might represent vibration signal acquired on a planetary gearbox used in a bucket wheel excavator. It consists of deterministic part represented by sine waves and additive Gaussian noise. When the gearbox is locally damaged, the noise is amplitude modulated by a pulse train. Usually, this so called "faulty signal" is not visible in the time domain—the deterministic signal has much more energy, thus it completely contaminates the amplitude modulated noise. An essential issue related to vibrations of the planetary gearbox operating in the bucket wheel excavator is a periodical frequency modulation of deterministic components that is closely related to cyclic operation of the machinery.

Figure 4 presents a part of raw time series representing vibrations of healthy and damaged planetary gearboxes. In fact, entire signals last 4 s. It consist of 4 sine waves of 500, 1000, 1500 and 2000 Hz with frequency modulated by a sawtooth wave with

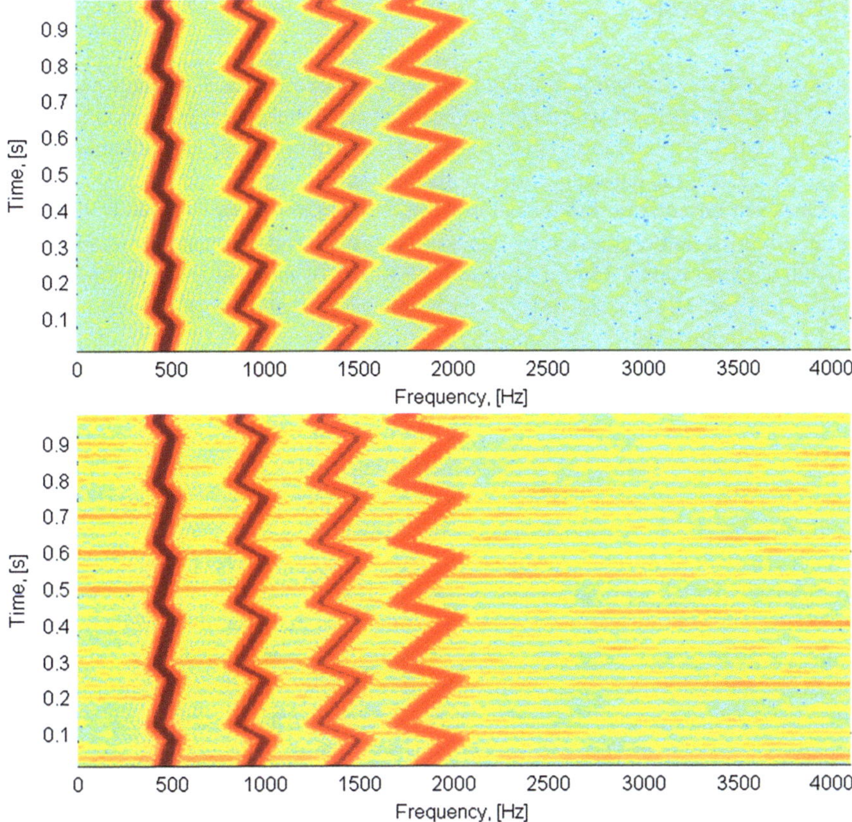

Fig. 5 Spectrograms of healthy (*top panel*) and faulty (*bottom panel*) signal. Note horizontal lines for faulty data

modulation frequency 6 Hz. In the case of damage the noise amplitude modulation equals to 30 Hz. As it can be seen, the pulse train is masked by deterministic components in the time domain.

A time-frequency representation of the signals is shown in Fig. 5. To preserve visibility of significant components only the first second of the signals is plotted. Wide band character of impulsive noise can be seen as horizontal lines. They occur only in the signal which represents the damaged gearbox.

The essential issue is to remove frequency modulated sine waves while preserving as much signal of interest as possible. Thus, filtering methods with frequency characteristics constant in time (e.g. filters based on AR model) might not be optimal.

4 Data Analysis

In order to show possibility of using PAR model to extract the SOI, order and period must be specified. As it was mentioned earlier, we assume that the period is known when frequency modulation is consisted with cyclic operation of the bucket wheel excavator. Here, the period is equal to 2048 samples.

Once our target is to remove deterministic components, we decided to fit appropriate order of autoregression by checking residuals for the presence of frequency modulated sine waves. Figure 6 presents residuals of the best fitting model, which order equals to 15 (note the order of PAR $<< T$ for considered machine).

In this chapter we do not incorporate best-fitting criteria (Akaike information criterion, Bayesian information criterion, final prediction error). We determine order of PAR experimentally and assume that a model is well-fitted if the residual time series imitates the initial noise. In analyzed cases, the residuals of PAR models with order 15 definitely look like the initials, especially in the faulty signal, where wideband excitations are visible in the whole spectrum (see Fig. 6, bottom panel).

Figure 7 presents envelope spectra for healthy and faulty signals presented in Fig. 6. The fault frequency of 30 Hz is clearly visible. There are no harmonics related to frequency which modulated sine waves.

The sample autocorrelation function of residuals' squares confirms fault frequency equal to 30 Hz which stands for period of about 273 samples (Fig. 8).

Recall that analyses in this section were performed for residuals obtained by fixing the periodical variance to 1, i.e. only periodical autoregression coefficients were used to obtain residuals. In Fig. 9 we present series of estimated periodical variance.

5 AR Versus PAR: Comparison of Residuals

To show indisputably superiority of PAR over ordinary autoregressive model (AR), with constant coefficients, we have fitted AR models with the same order as PAR to our simulated signals.

Figure 10 shows time series of residuals. In both cases the series do not look like set of independent random variables. Spectrograms of these signals clearly illustrate nature of this behavior (Fig. 11). Energy of sine waves is minimized, but they still remain in residuals. Looking at envelope spectra of residuals obtained by PAR (Fig. 7) and AR (Fig. 12) it is clear that residuals obtained using the model with time-varying coefficients is much more effective.

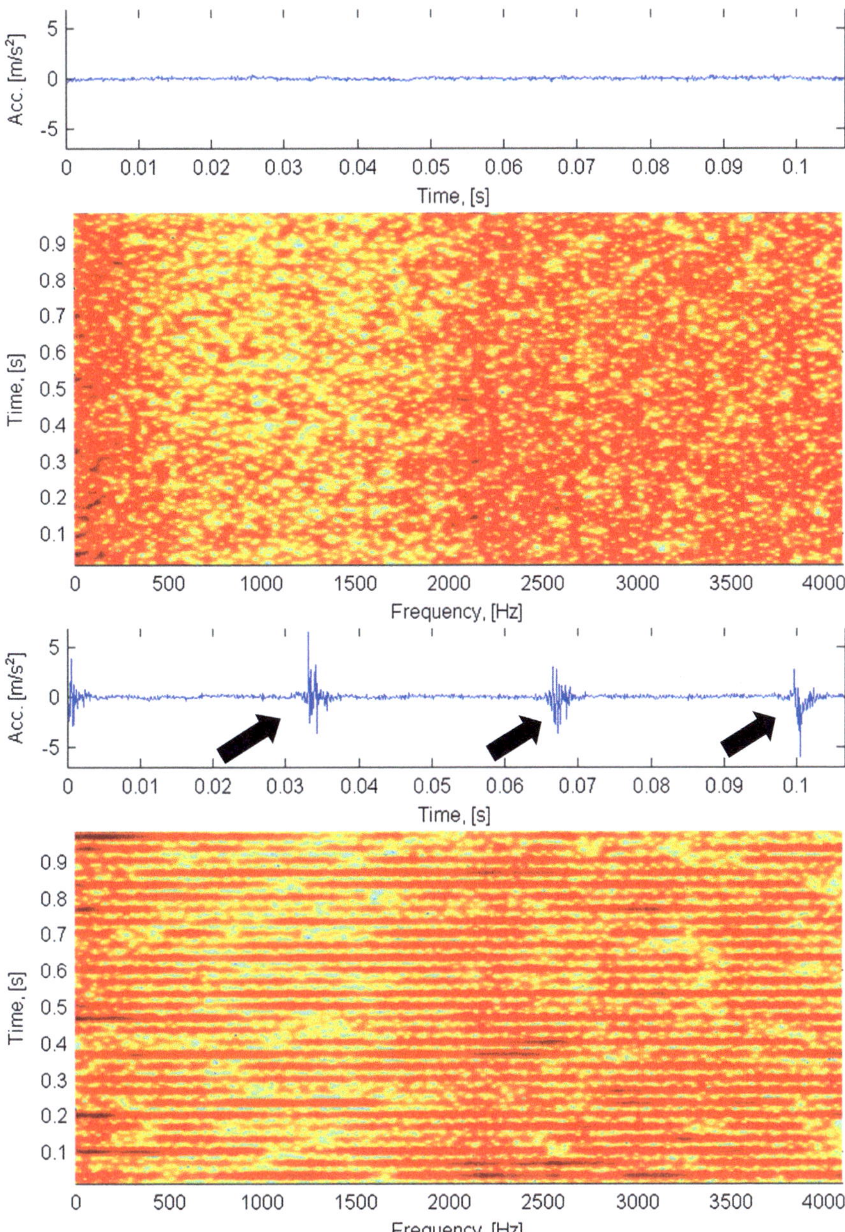

Fig. 6 Filtered signals using PAR with order equal to 15. Healthy (*first and second panel*) and faulty (*third and fourth panel*). Observe deterministic components completely removed and wide-band excitations visible in the whole spectrum

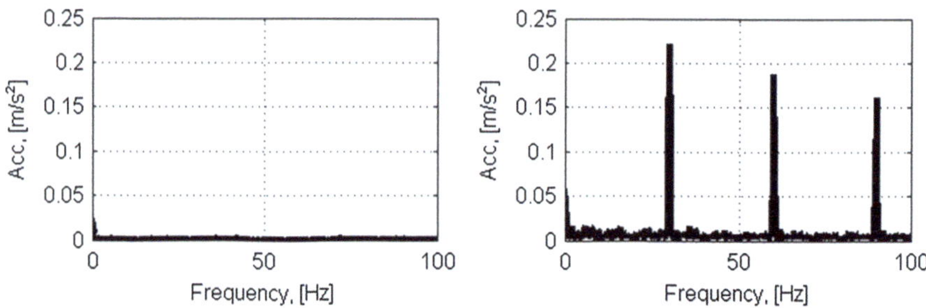

Fig. 7 Envelope spectra for filtered signal (PAR order = 15). Healthy (*left panel*) and faulty (*right panel*). Peaks on the right panel correspond to fault frequency—30 Hz

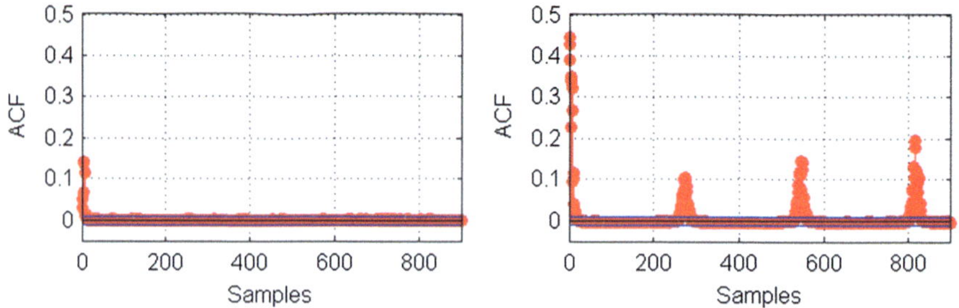

Fig. 8 Sample autocorrelation function for squared filtered signal (PAR order = 15). Healthy (*left panel*) and faulty (*right panel*). Peaks on the right panel correspond to fault frequency—30 Hz (about 273 samples)

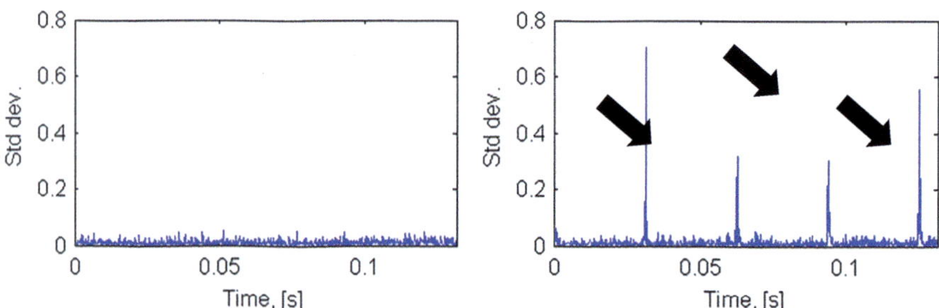

Fig. 9 Estimated periodical standard deviation for healthy (*left panel*) and faulty (*right panel*) signal. Peaks on the right panel correspond to 1/(fault frequency)

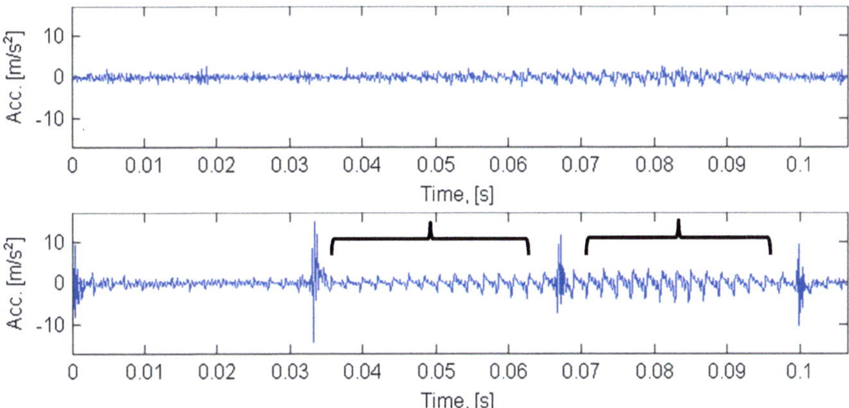

Fig. 10 Filtered signals using AR with order equal to 15. Healthy (*top panel*) and faulty (*bottom panel*). Observe not noise-looking shape of the signal between distributions

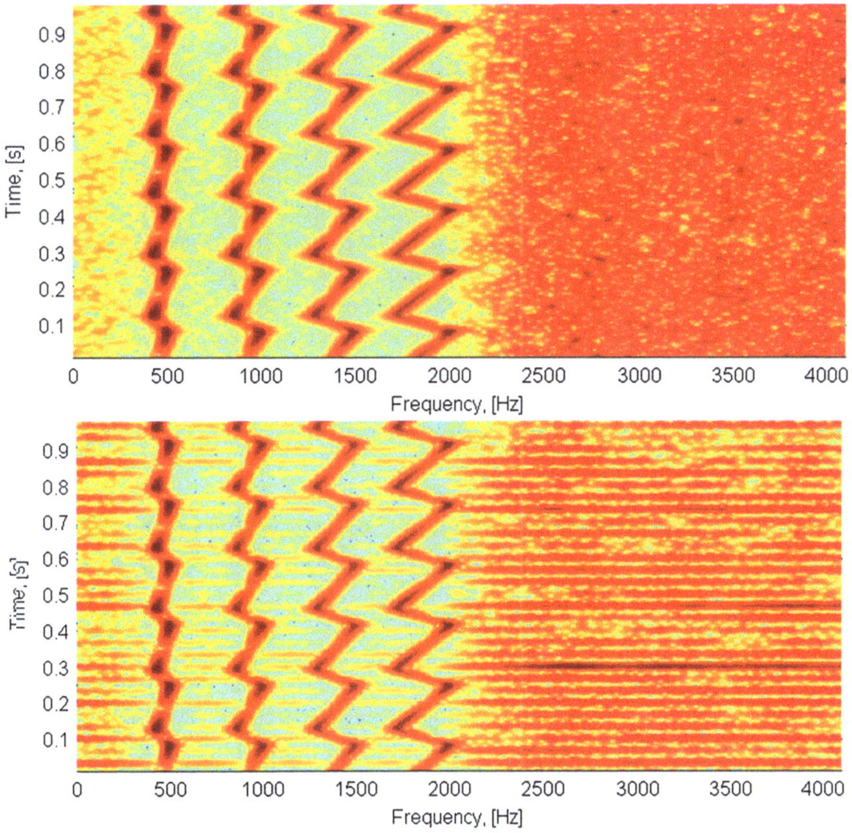

Fig. 11 Spectrograms of filtered signals using AR with order equal to 15. Healthy (*top panel*) and faulty (*bottom panel*). Observe deterministic components barely removed

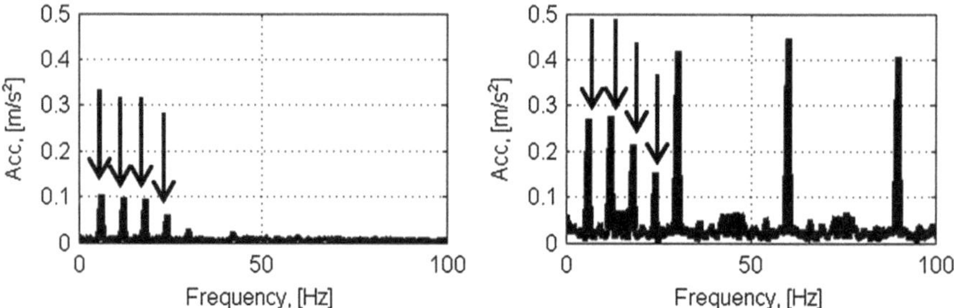

Fig. 12 Envelope spectra of filtered signals using AR with order equal to 15. Healthy (*left panel*) and faulty (*right panel*). Observe harmonics related to 6 Hz frequency modulation of the remains of deterministic components

6 Conclusions

In this chapter a novel procedure for modeling and processing of vibration signals focused on damage detection in rotating machinery is proposed. It can be considered as an extension of known AR approach directed to non-stationary signals with time varying (cyclic) frequency structure.

To illustrate core of the problem and basis of our solution we refer to vibration data from bucket wheel excavator, heavy duty machine used in mining industry. During its operation, the machine is subjected to time varying, cyclic (nearly periodic) variation of external load. It was shown, that varying load causes varying spectral structure of vibration generated by planetary gearbox incorporated in drive unit of BWE.

The task for diagnostics is to detect local damage in the gearbox using vibration signal. Model-based signal processing used in such context was taking advantage of autoregressive modeling. It is well known, that AR modeling is useful for describing discrete frequency structure of the signal and from mechanical perspective can match to shaft, mesh and other predefined frequencies. Local damage in rotating machine produces impulsive, cyclic excitation that is equivalent to wideband and cyclic excitation in frequency domain. Such wideband signature is nearly invisible for AR model (no information about discrete frequency, no information about localization in time). Using AR model one might design inverse filter that "dump" these discrete frequencies and "pass" rest of the signal.

In case of time varying external load, when spectral content follows variation of load, using AR with constant coefficients is not reasonable.

AR model will try to match to averaged frequency instead of tracking its instantaneous variation. Due to cyclic variation of spectral content, it is proposed to use periodic AR (PAR). In the chapter we formulated the problem, discussed estimation issues, applied procedure to the data and finally compared to AR with constant coefficients.

To validate results we used both "healthy" and "faulty" data, comparison covers analysis of spectrograms (visual inspection) and three other techniques: envelope spectra, sample autocorrelation function and periodical standard deviation—all calculated for PAR-based filtered data for healthy and faulty cases.

Further work will focus on validation proposed technique for real data. Here, to illustrate the idea we used synthetic signal (simulations) only.

References

Antoni J (2005) Blind separation of vibration components: principles and demonstrations. Mech Syst Signal Pr 19:1166–1180

Antoni J, Bonnardot F, Raad A, El Badaoui M (2004) Cyclostationary modeling of rotating machine vibration signals. Mech Syst Signal Pr 18:1285–1314

Baillie DC, Mathew J (1996) A comparison of autoregressive modeling techniques for fault diagnosis of rolling element bearings. Mech Syst Signal Pr 10:1–17

Bartelmus W, Zimroz R (2009) Vibration condition monitoring of planetary gearbox under varying external load. Mech Syst Signal Pr 23:246–257

Bloomfield P, Hurd HL, Lund RB (1994) Periodic correlation in stratospheric ozone data. J Time Ser Anal 12:127–150

Brockwell PJ, Davis RA (2006) Time Series: Theory and Methods. Springer Series in Statistics, Springer, New York

Broszkiewicz-Suwaj E, Makagon A, WeronR Wylomanska A (2004) On detecting and modeling periodic correlation in financial data. Physica A 336:196–205

Chaari F, Bartelmus W, ZimrozR, Fakhfakh T, Haddar (2012) Gearbox vibration signal amplitude and frequency modulation. Shock Vibr 19:635–652

Combet F, Zimroz R (2009) A new method for the estimation of the instantaneous speed relative fluctuation in a vibration signal based on the short time scale transform. Mech Syst Signal Pr 23:1382–1397

Endo H, Randall RB (2007) Enhancement of autoregressive model based gear tooth fault detection technique by the use of minimum entropy deconvolution filter. Mech Syst Signal Pr 21:906–919

Gardner W, Franks LE (1975) Characterisation of cyclostationary random signal processes. IEEE T Inform Theory 21:4–14

Gladyshev EG (1961) Periodically correlated random sequences. Sov Math 2:385–388

Hurd HL, Miamee A (2007) Periodically Correlated Random Sequences. Spectral Theory and Practice, Wiley, New Jersey

Makowski RA, Zimroz R (2011) Adaptive bearings vibration modelling for diagnosis. Lect Notes comput Sc (including subseries Lecture Notes in Artificial Intelligence and Lecture Notes in Bioinformatics) 6943:248–259

Makowski RA, Zimroz R (2013) A procedure for weighted summation of the derivatives of reflection coefficients in adaptive Schur filter with application to fault detection in rolling element bearings. Mech Syst Signal Pr 38:65–77

Nowicka-Zagrajek J, Wyłomańska A (2006) The dependence structure for PARMA models with a-stable innovations. Acta Phys Pol B 37:3071–3081

Obuchowski J, Wylomanska A, Zimroz R (2013) Stochastic modeling of time series with application to local damage detection in rotating machinery. Key Eng Mat 569:441–449

Parzen E, Pagano M (1979) An approach to modeling seasonally stationary time-series. J Econom 9:137–153

Poulimenos AG, Fassois SD (2006) Parametric time-domain methods for non-stationary random vibration modeling and analysis—a critical survey and comparison. Mech Syst Signal Pr 20:763–816

Randall RB, Antoni J (2011) Rolling element bearing diagnostics—a tutorial. Mech Syst Signal Pr 25:485–520

Samuel PD, Pines DJ (2005) A review of vibration-based techniques for helicopter transmission diagnostics. J Sound Vib 282:475–508

Urbanek J, Barszcz T, Antoni J (2013) Time-frequency approach to extraction of selected second-order cyclostationary vibration components for varying operational conditions. Measurement 46:1454–1463

Vecchia AV (1985) Periodic autoregressive-moving average (PARMA) modeling with applications to water resources. Water Resour Bull 21:21–730

Wang W, Wong AK (2002) Autoregressive model-based gear fault diagnosis. J Vib Acoust 124:172–179

Zhan Y, Makis V, Jardine AKS (2006) Adaptive state detection of gearboxes under varying load conditions based on parametric modeling. Mech Syst Signal Pr 20:188–221

Zhan Y, Mechefske CK (2007) Robust detection of gearbox deterioration using compromised autoregressive modeling and Kolmogorov-Smirnov test statistic-Part I: compromised autoregressive modeling with the aid of hypothesis tests and simulation analysis. Mech Syst Signal Pr 21:1953–1982

Zimroz R, Bartelmus W (2012) Application of adaptive filtering for weak impulsive signal recovery for bearings local damage detection in complex mining mechanical systems working under condition of varying load. Diffus Def B 180:250–257

Zimroz R, Urbanek J, Barszcz T, Bartelmus W, Millioz F, Martin N (2011) Measurement of instantaneous shaft speed by advanced vibration signal processing—application to wind turbine gearbox. Metrol Meas Sys 18:701–712

MIX
Papier aus verantwortungsvollen Quellen
Paper from responsible sources
FSC® C105338

If you have any concerns about our products,
you can contact us on
ProductSafety@springernature.com

In case Publisher is established outside the EU,
the EU authorized representative is:
Springer Nature Customer Service Center GmbH
Europaplatz 3, 69115 Heidelberg, Germany

Printed by Libri Plureos GmbH
in Hamburg, Germany